SIGNAL PROCESSING OF RANDOM PHYSIOLOGICAL SIGNALS

Signal Processing of Random Physiological Signals

Charles S. Lessard

ISBN: 978-3-031-00482-7   paper   Lessard
ISBN: 978-3-031-01610-3   ebook   Lessard

DOI: 10.1007/978-3-031-01610-3

A Publication in the Springer series
SYNTHESIS LECTURES ON BIOMEDICAL ENGINEERING
Lecture #1
Library of Congress Cataloging-in-Publication Data

First Edition

SIGNAL PROCESSING OF RANDOM PHYSIOLOGICAL SIGNALS

Charles S. Lessard, Ph.D.
Texas A&M University, College Station, USA

SYNTHESIS LECTURES ON BIOMEDICAL ENGINEERING #1

ABSTRACT

This *lecture* book is intended to be an accessible and comprehensive introduction to random signal processing with an emphasis on the real-world applications of *biosignals*. Although the material has been written and developed primarily for advanced undergraduate biomedical engineering students it will also be of interest to engineers and interested biomedical professionals of any discipline seeking an introduction to the field. Within education, most biomedical engineering programs are aimed to provide the knowledge required of a graduate student while undergraduate programs are geared toward designing circuits and of evaluating only the cardiac signals. Very few programs teach the processes with which to evaluate brainwave, sleep, respiratory sounds, heart valve sounds, electromyograms, electro-oculograms, or random signals acquired from the body. The primary goal of this *lecture* book is to help the reader understand the time and frequency domain processes which may be used and to evaluate random physiological signals. A secondary goal is to learn the evaluation of actual mammalian data without spending most the time writing software programs. This publication utilizes "DADiSP", a digital signal processing software, from the DSP Development Corporation.

KEYWORDS

Signals, Processes, Time Domain, Frequency Domain, Random Data

Contents

C H A P T E R   1

Biomedical Engineering Signal Analysis

1.1   INTRODUCTION

The purpose of this book is to present the most widely used techniques in signal and system analysis. Individuals should have sufficient working knowledge of mathematics through calculus and some physiology and be familiar with the elements of circuit theory (i.e., you can write both loop and node equations for passive and active circuits) to gain the most knowledge. Extensive programming ability is not necessary if an individual wishes to apply some of the signal-processing principles, and it is recommended that the individual link to DSP Corporation web site (http://www.dadisp.com/) and try the Student version of the digital signal-processing software package called "DADiSP." The material contained herein should serve as an introduction to the most widely used techniques in analog and discrete (digital) data and signal-processing analysis of physiological data.

1.2   GENERALIZED SYSTEMS ENGINEERING APPROACH

This chapter is concerned with methods of characterizing signals and systems. Systems can be studied from two main perspectives:

a)   *Microscopic systems analysis approach*: The fine structure of a system is taken into account. This approach is an extremely difficult to analyze because of the complexity and large number of variables in the mathematical description of the system. Muscles are a good example. A model of a muscle must consider its components (fibrin, actin, myosin, and the level of action potential) in building up the comprehensive description of the system operation.

b) *Macroscopic system analysis method*: This method is the most common and most useful approach in system analysis. In this approach, the system is characterized in terms of subsystems and components (usually lumped parameter components), which is the analysis that we will be primarily concerned with.

Macroscopic analysis requires that the system be broken into a number of individual components. The various components are described in sufficient detail and in a manner so that the system operation can be predicted. The crux is the description of the component behavior, which is done in terms of a mathematical model.

For many components of engineering interest, the mathematical model is an expression of the response of the component (subsystem) to some type of stimulus or forcing function. In our mathematical models, the forcing function and the system response are variables, which we will categorize as signals. Signal or data (variables) could be displacement, voltage, price of stock, numbers in an inventory, etc.

As stated before, the purpose of an engineering analysis of a system is to determine the response of the system to some input signal or excitation. Response studies are used to:

1) establish performance specifications;

2) aide in selection of components;

3) uncover system deficiencies (i.e., instabilities);

4) explain unusual or unexplained behavior;

5) establish proper direction and ranges of variables for the experimental program. The final system design requires a combined analytical and experimental approach; and

6) analytical studies are invaluable in the interpretation of results.

Many systems cannot be analyzed as a whole because of the enormous complexity of the systems and the lack of a satisfactory mathematical model or models for the systemwide component; for example, large systems such as an aircraft, spacecraft, etc. Subsystems can be treated and studied independently: the body, brain, central nervous system, endocrine system, etc.

To study and analyze a system properly, the means by which energy is propagated through the system must be studied. Evaluation of energy within a system is done by specifying how varying qualities change as a function of time within the system. A varying quantity is referred to as a *signal*. Signals measure the excitation and responses of systems and are indispensable in describing the interaction among various components/subsystems. Complicated systems have multi-inputs and multioutputs, which do not necessarily have to be of the same number. Signals also carry information and coding (i.e., encoded messages), which has led to the field of signal theory in communication systems. Systems analysis is used to find the response to a specific input or range of inputs when

1) the system does not exist and is possible only as a mathematical model;

2) experimental evaluation of a system is more difficult and expensive than analytical studies (i.e., ejection from an airplane, automobile crash, forces on lower back); and

3) study of systems under conditions too dangerous for actual experimentation (i.e., a/c ejection, severe weather conditions).

System representation is performed by means of specifying relationships among the systems variables, which can be given in various forms, for example, graphs, tabular values, differential equations, difference equations, or combinations. We will be concerned primarily with the system representation in terms of ordinary linear differential equations with constant coefficients.

There are two main questions one needs to ask:

1. What is the appropriate model for a particular system?

2. How good a representation does the model provide?

Often how well a model represents the real system is measured through the "squared correlation coefficient" or its equivalent the "coherence function," which are often referred to as "The Measure of Goodness Fit." You will learn more about these subjects in the later chapters.

CHAPTER 2

System Classification

As mentioned in Chapter 1, a system as defined in engineering terms is "a collection of objects interacting with each other to accomplish some specific purpose." Thus, engineers tend to group or classify systems to achieve a better understanding of their excitation, response, and interactions among other system components. It should first be noted that the general conditions for any physical system can be reflected by the following general model equation (2.1):

$$A_n(t)\frac{d^n y}{dt^n} + A_{n-1}(t)\frac{d^{n-1}y}{dt^{n-1}} + \cdots + A_1(t)\frac{dy}{dt} + A_o(t)y$$

$$= B_m(t)\frac{d^m x}{dt^m} + B_{m-1}(t)\frac{d^{m-1}x}{dt^{m-1}} + \cdots + B_1(t)\frac{dx}{dt} + B_o(t)x \qquad (2.1)$$

Thus, the necessary conditions for any physical system are (1) that for an excitation (input) the function $x(t) = f(\phi)$ exists for all $t < t_o$ and (2) that the corresponding response (output) function $y(t) = f(\phi)$ must also exist for all $t < t_o$. These conditions are important because natural physical system cannot anticipate an excitation and responds before the excitation is applied.

Classifying a signal typically involves answering several questions about the system, as shown in Table 2.1.

Let us begin with definitions of terms used to characterize systems.

2.1 ORDER

What is the order of a system? The order of a system is the highest order derivative. Determining the order of a system is important because it also characterizes the response

TABLE 2.1: How Are Systems Classified
When classifying a system, the following questions should be considered: 1. What is the order of the system? 2. Is it a causal or noncausal system? 3. Is it a linear or nonlinear system? 4. Is it a fixed or time-varying system? 5. Is it a lumped or distributed parameter system? 6. Is it a continuous or discrete time system? 7. Is it an instantaneous or a dynamic system?

of the system. For example, (2.2) may be used to represent a system as a differential equation and to classify the system as an "nth" order system.

$$a_n \frac{d^n y}{dt^n} + a_{n-1} \frac{d^{n-1} y}{dt^{n-1}} + \cdots + a_1 \frac{dy}{dt} + a_\circ (t) \, y$$
$$= b_m \frac{d^m x}{dt^m} + b_{m-1} \frac{d^{m-1} x}{dt^{m-1}} + \cdots + b_1 \frac{dx}{dt} + b_\circ y \qquad (2.2)$$

2.2   CAUSAL VS. NONCAUSAL

What is a causal or noncausal system? A "Causal System" is defined as a physical nonanticipatory system whose present response does not depend on future values of the input, whereas a "Non-causal System" does not exist in the real world in any natural system. "Non-causal System" exists only when it is man-made by simulating the system with a time delay.

2.3   LINEAR VS. NONLINEAR

What is a linear or a nonlinear system? From previous classes you have learned that when the output vs. the input is a straight line, the system is considered to be linear. Do not use this definition. Let us extend the simple straight-line definition to a more general analytical or mathematical expression. By mathematical definition, "A linear system is a system that possesses the mathematical properties of associative, commutative, and distributive."

Other key facts noteworthy of consideration regarding limitations in determining linearity include the following.

1. *Superposition* applies for linear systems *only*, but superposition can be applied only to linear systems that possess the mathematical properties of associative, commutative, and distributive.

2. It should also be noted that in linear systems components do not change their characteristics as a function of the magnitude of excitation. For example, electical components R, L, and C do not change their values because of changes in the magnitude of voltage or current running through them, or in the time period of analysis.

3. Some components are linear within a range.

That a system is linear can be shown by proving that superposition applies. Note that superposition is expressed as input output, the following equations are applicable:

If $y_1(t)$ is the system response to $x_1(t)$, and

If $y_2(t)$ is the system response to $x_2(t)$, then

if the inputs are summed, the output of the system should yield the sum of the individual responses as shown in (2.3): Associative property,

$$y_1(t) + y_2(t) = x_1(t) + x_2(t) \qquad (2.3)$$

Or, if the responses (outputs) of two weighted inputs are given in Eq. 2.4.

$$y_1(t) = ax_1(t)$$
$$y_2(t) = bx_2(t) \qquad (2.4)$$

Then, if the inputs are summed, the response (output) of the system should yield the sum of the individual responses (2.5) such that

$$y_1 + y_2 = ax_1 + bx_2 \quad \text{for all} \quad a, b, x_1(t), x_2(t) \tag{2.5}$$

A simple, but not an accurate way to define a linear system is as follows: "A system in which all derivatives of the excitation and the response are raised to the 1st power."

A nonlinear system is a system in which derivatives of excitation and response are greater than the 1st order (power). Nonlinear systems or components are often modeled as 2nd or higher order systems, that is, rectifiers, junction diodes, and nonlinear magnetic devices.

2.4   FIXED VS. TIME-VARYING

What is a fixed or time-varying system? A fixed system means that coefficients do *not* vary with time, whereas a time-varying system means that coefficients do vary with time.

Another way of thinking of this definition is that if the excitation, $x(t)$, results in a response, $y(t)$, then the excitation at some time later, $x(t - \tau)$, will result in a response, $y(t - \tau)$, for any excitation $x(t)$ and any delay τ.

As an example, in (2.1), it can be noted that the system is time varying, since the coefficients $a_i(t)$ and $b_i(t)$ are indicated as being functions of time; however, in Eq. 2.2, the coefficients a_i and b_i are not functions of time, hence the system is a "Fixed System."

Analysis of time-varying systems is very difficult; thus, this book will concentrate on differential equations with constant coefficients such that the system is said to be fixed, time-invariant, or stationary. One can add that for "Time Invariant Systems," the slope of the system response depends on the shape of the system excitation and not on the time when the excitation was applied.

2.5   LUMPED PARAMETER VS. DISTRIBUTED PARAMETER

What is a lumped or distributed parameter system? A lumped parameter system is defined as a system whose largest physical dimension is small compared to the wavelength of the highest significant frequency of interest. Equation (2.1) is representative of a lumped parameter system by virtue of being an ordinary differential equation. It implies that the physical size is of no concern, since the excitation propagates through the system instantaneously, which is valid only if the largest physical dimension of the system is small compared to the wavelength of the highest significant frequency considered.

A distributed parameter system is defined as a system whose dimensions are large (*NOT Small*) compared to the shortest wavelength of interest. Distributed-parameter systems are generally represented by partial differential equations. For example, waveguides, microwave tubes, and transmission lines (telephone and power lines) are all distributed parameter systems, because of their physical lengths; that is, Hoover Dam to Los Angeles are much larger than the highest frequency or shortest wavelength of interest.

Note: As the frequency increases (higher frequency), the wavelength decreases.

2.6 CONTINUOUS-TIME VS. DISCRETE TIME

What is a Continuous or Discrete time system? "Continuous-Time Systems" are those that can be represented by continuous data or differential equations, whereas "Discrete Time Systems" are represented by sampled digital data.

It should be noted that all natural physical systems are inherently continuous-time, since time is continuous. In addition, it should be noted that both Continuous- and Discrete time systems can also be

1. linear or nonlinear,
2. fixed or time varying, and
3. lumped or distributed.

2.7 INSTANTANEOUS VS. DYNAMIC

What is an Instantaneous or a Dynamic System? An Instantaneous system is defined as a system that has no memory, which means that the system is not dependent on any future or past value of excitation. The response at some point in time (t_1) of an Instantaneous system depends only on the excitation at time (t_1). For example, a fully resistive circuit has no memory elements, that is, stored charge in a capacitor, and is considered an instantaneous system. There is no need to express the system as a differential equation, only as constants.

If a system response depends on past value of excitation, then the system is classified as a Dynamic System, which has memory (meaning it stores energy). Circuits with capacitors and magnetic inductors components are Dynamic systems, since the capacitors initial state is based on past excitation. Similarly, magnetic components with residual magnetism store energy based on past excitation, thus making such systems "Dynamic." Dynamic system mathematical models are expressed as differential equations in either time or frequency domain.

In general, most problems are considered initially to be

1. causal,

2. linear,

3. fixed,

4. lumped-parameter,

5. continuous time, and

6. dynamic systems.

CHAPTER 3

Classification of Signals

In this chapter, an overview of various methods to describe signals in a quantitative manner is presented. Most medical devices in the 1950s and 60s presented their output in a strip chart for the physician to view and make some determination (diagnosis); however, with the advent of digital computers and advanced automated analytical processes, it has become necessary to classify and quantify signals.

So, the questions are: "Why is it necessary to quantify signals?" and "How is it useful?" Reasons for signal quantification are as follows:

1. By representing signals as mathematical models, the engineer/researcher is able to carry out system analyses under specified conditions

2. The engineer may be after the information that the signal carries.
    a. Can your thoughts be decoded from your brainwaves?

3. To specify the form of a signal that will be used in the design of systems to perform specific tasks.
    a. Examples:
        i. Petrofski's work in feeding electrical signals back into paralytic limbs in order to cause the patient to walk
        ii. Auditory or visual prostheses?

4. Finally, classification and quantification provide physical insight for the analyst or designer.
    a. If an engineer wants to design a "Star Trek Tricoder" for vital life signs, what signal characteristics and values does the engineer need to consider?

There are some useful approximations of a signal that can be used to quantify the signal such that the signal is represented in some mathematically precise format. The most general representation of a signal, which is a function of time, is the abstract symbol $x(t)$ or $y(t)$. These abstract symbols can be used with linear systems because the system characteristics are independent of the explicit forms of the signals. But also note that the abstract representation, $x(t)$, is **NOT** a quantitative representation, since the abstract notation does not specify the magnitude of the signal at any instant in time.

To specify the magnitude at any instant of time, the signal must be specified more explicitly by a function of time defined for all instants of time. Realistically, any quantitative representation is only an approximation. So that leaves the engineer with the next problem, which is that of deciding how good the approximation is in order to obtain meaningful results from its use.

3.1   HOW ARE SIGNALS CLASSIFIED

Let us begin to examine the various ways that signals are classified. In this section, only the most used classifications are presented. Engineers should ask the following questions (see Table 3.1).

TABLE 3.1: Signal Classification
1.   Is the signal periodic or nonperiodic?
2.   Is the signal random or nonrandom (deterministic)?
3.   Is the signal an energy signal or a power signal?
4.   What is the bandwidth of the signal?
5.   What is the time duration of the signal?
6.   What is time-bandwidth product?
7.   What is the dimensionality of the signal?

3.1.1 Periodic vs. Nonperiodic Signals

Is the signal periodic or nonperiodic? A **periodic** signal is defined as a signal that repeats the sequences of values exactly after a fixed length of time (the period, T) as shown in (3.1), such that

$$x(t) = x(t + T) \quad \text{for all} \quad t \tag{3.1}$$

Keep in mind that a delay is written as a difference $(t - T)$ in (3.2):

$$y(t) = x(t - T) \tag{3.2}$$

The smallest positive T, which satisfies 3.1, is the period, which is the duration of one complete cycle. The fundamental frequency (f_0) of a periodic signal is defined as the reciprocal of the period (T) as in (3.3):

$$f_0 = \frac{1}{T} \tag{3.3}$$

A **nonperiodic** signal or an "aperiodic" (almost periodic) signal does not have a period (T). Typical nonperiodic signals include signals such as speech, electroencephalograms (brainwaves) measured by surface electrodes, electromyogram, and respiratory sounds.

From the standpoint of mathematical representation of signals, the periodic class has the greatest theoretical importance. In most cases, an explicit mathematical expression can be written for a periodic signal.

3.1.2 Random vs. Nonrandom Signals

Is the signal random or nonrandom (deterministic)? Another method of classifying signals is whether or not the signal is random. The author's definition of a **random** signal is a signal about which there is some degree of uncertainty before it actually occurs; that is, "I've always considered a random signal as a time varying variable or function whose magnitude varies erratically and in an unpredictable manner."

The formal definition of a **nonrandom** signal is a signal that has no uncertainty before it occurs. In other words, it is "**Deterministic**"; in most cases, an explicit math expression can be written.

3.1.3   Energy Signals vs. Power Signals

Is the signal an energy signal or a power signal? Since we are most familiar with electrical measurements, let us recall that most electrical signals are voltages or currents. The energy dissipated by a voltage in a resistor for a given period of time is given by (3.4).

$$E = \int_{t_1}^{t_2} \frac{v^2(t)}{R} dt \quad \text{watt seconds} \tag{3.4}$$

Similarly for current, the equation is given in (3.5):

$$E = \int_{t_1}^{t_2} R i^2 dt \tag{3.5}$$

In either equation, the energy is proportional to the integral of the square of the signal. If the resistance (R) is made unity (R is set at 1 ohm: 1Ω), then (3.4) and (3.5) may be written in the more generalized form as shown in (3.6).

$$E = \int_{-\infty}^{+\infty} x^2(t) dt < \infty \tag{3.6}$$

Thus (3.4) is used to define an energy signal of infinite energy. An exponentially damped sinusoid is an example of an energy signal.

What is a power signal? **Average power** is also known as a "Power Signal." If the resistance is assumed to be 1Ω, then the equation for a power signal is given by (3.7).

$$P = \frac{1}{t_2 - t_1} \int_{t_1}^{t_2} x^2(t) dt \tag{3.7}$$

From (3.7), one may conclude that the "Power" signal is the energy of a signal in a time interval divided by the time interval, or restated: Power is the average energy in an interval of time.

Average power (3.8) satisfies the following conditions: [Note the limits; it is bound.]

$$0 < \lim_{T \to \infty} \frac{1}{2T} \int_{-T}^{T} x^2(t)dt < \infty \tag{3.8}$$

A classic definition given by McGillem and Cooper is as follows:

> If the **power signal** has a non-zero value when the time interval becomes infinite, then the signal has a finite average power and is called a **Power Signal**. Then, an **energy signal** has zero average power, whereas a **power signal** has infinite energy.

3.2 SIGNAL CHARACTERIZATION (MEASURABLE PARAMETERS)

Thus far, we have simply provided discussions on signal representation. Let us turn to specification of the signals and the "Characterization" of the signals by a few significantly useful parameters, which include the following parameters:

1. Signal energy

2. Average power

3. Bandwidth, or

4. Time duration

5. Time–bandwidth product

6. Dimensionality of the signal

The first two, Energy and Power, were treated in the previous section. Thus, let us start with the question, "What is the time duration of a signal?" The "Time Duration" of a signal is defined in statistical terms as the normalized variance about its central moment where the 1st moment is the mean or its center of gravity and the variance is generally the second moment. The definition is stated in a statistical sense, which is especially useful for signals that do not have a definite starting or stopping time.

In the case of the Time Duration of a signal, the moments are normalized by dividing the moments by the energy associated with the signal [the generalized energy equation (3.9)]

$$E = \int_{-\infty}^{\infty} x^2 dt \tag{3.9}$$

Then, the normalized first moment is shown in (3.10):

$$t_0 = \frac{\int_{-\infty}^{\infty} t x^2(t) dt}{\int_{-\infty}^{\infty} x^2(t) dt} \tag{3.10}$$

And the normalized second-central moment of the square of the time function is given by (3.11):

$$T_e = \left[\frac{\int_{-\infty}^{\infty} (t - t_0)^2 \, x^2(t) dt}{\int_{-\infty}^{\infty} x^2(t) dt} \right]^{\frac{1}{2}} \tag{3.11}$$

where T_e defines the measure of the time duration, and t_0 is the normalized first moment or mean about the origin. Therefore, the time duration (T_e) is in reality the normalized **second** moment about the mean, t_0.

3.2.1 Bandwidth

What is the bandwidth of the signal? Most engineers are very familiar with the definition of bandwidth, but let us review the various definitions. Bandwidth is intended to indicate a range of frequencies within which most of the energy lies. Engineers use the half power or 3 dB point, if a frequency response plot is available (Fig. 3.1).

However, the bandwidth of a general signal is defined in terms of the derivative of the time function as shown in (3.12).

$$W_e = \frac{1}{2\pi} \left[\frac{\int_{-\infty}^{\infty} \left(\frac{dx(t)}{dt} \right)^2 dt}{\int_{-\infty}^{\infty} x^2(t) dt} \right]^{\frac{1}{2}} \text{Hz} \tag{3.12}$$

Caution: This definition has problems with discontinuities in the signals.

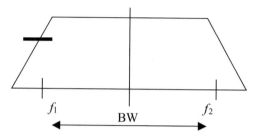

FIGURE 3.1: Frequency response plot. The 3 dB or 0.707 magnitude point shown in bold horizontal line determines the low- and high-frequency cutoff points

3.2.2 Time–Bandwidth Product

What is time–bandwidth product? Another concept in characterizing signals is time–bandwidth product, which is the product $(T_e W_e)$ of the signal time duration and the signal bandwidth as given in (3.13).

$$TW = (T_e)(W_e) = \frac{\int_{-\infty}^{\infty} (t - t_0)^2 x^2(t)dt}{\int_{-\infty}^{\infty} x^2(t)dt} \frac{1}{2\pi} \left[\frac{\int_{-\infty}^{\infty} \left(\frac{dx(t)}{dt}\right)^2 dt}{\int_{-\infty}^{\infty} x^2(t)dt} \right]^{\frac{1}{2}} \qquad (3.13)$$

No signal can have a time–bandwidth product smaller than $\frac{1}{4\pi}$. Small values of the time–bandwidth product are associated with very simple signals. Large values imply complex structure and large information content. The time–bandwidth product is a parameter for judging the usefulness of a given signal in conveying information.

3.2.3 Dimensionality

What is the dimensionality of a signal? Signal dimensionality is a concept closely related to time–bandwidth product. Dimensionality of a signal is the smallest number of basis functions needed to achieve a reasonable approximation with a finite number of coefficients. Signal dimensionality is defined as twice the time–bandwidth product plus one. The equation for the dimensionality of a signal is given in (3.14).

$$D \cong 2WT + 1 \qquad (3.14)$$

where W is the bandwidth of the signal, and T the time duration of the signal.

3.3   REFERENCE

McGillem, C.D. and G.R. Cooper, "Continuous and Discrete Signal and System Analysis", Holt, Rinehart, and Winston, Inc., 1974.

CHAPTER 4

Basis Functions and Signal Representation

4.1 INTRODUCTION TO BASIS FUNCTIONS

To obtain a quantitative description of a signal, it is necessary to represent the signal in terms of explicit time functions. Mathematical convenience dictates that the signal can be represented as a linear combination of a set of elementary time functions called **Basis Functions**. Equation (4.1) is the general equation representing a signal, $x(t)$, in terms of a set of basis functions designated as $\Phi_0(t)$, $\Phi_1(t)$, $\Phi_2(t)$, ..., $\Phi_N(t)$.

$$x(t) = \sum_{n=0}^{N} a_n \Phi_n(t) \tag{4.1}$$

4.2 DESIRABLE PROPERTIES OF BASIS FUNCTIONS

One property that is desirable for a set of basis functions is the *finality of coefficients*, which allows the determination of a coefficient without needing to know any other coefficients. This means that more terms can be added without making changes or recalculating the earlier or previous coefficient. To achieve "finality of coefficients," the basis functions must be orthogonal over the time interval of interest.

The condition of orthogonality requires that the integral of two functions satisfies two conditions. The first condition is when the two basis functions are not equal, then the resulting integral must be zero as shown in (4.2).

$$\int_{t_1}^{t_2} \Phi_n(t)\Phi_k(t)dt = 0; \qquad k \neq n \tag{4.2}$$

The second condition is when the two basis functions are equal, then the resulting integral will equal some value "Lambda," λ_k, as shown in (4.3).

$$\int_{t_1}^{t_2} \Phi_n(t)\Phi_k(t)dt = \lambda_k; \; k = n \quad \text{for all } k \text{ and } n \tag{4.3}$$

If both sides of (4.3) are divided by λ_k and lambda λ_k is made unity ($\lambda_k = 1$) for all k, then the basis function is called **orthonormal**. Note that the limits of integration can be defined as finite interval or an infinite or semi-infinite interval.

4.3 EVALUATION OF COEFFICIENTS

Recall the general expression of a signal represented in terms of a basis function, as the weighted summation of the elementary functions:

$$x(t) = \sum_{n=0}^{N} a_n \Phi_n(t) \tag{4.4}$$

Hence, to evaluate the coefficients a_n of the basis function, the following steps are necessary:

(1) Multiply both sides by $\Phi_j(t)$ as in (4.5).

$$\Phi_j(t)x(t) = \Phi(t)\sum_{n=0}^{N} a_n \Phi_n(t) \tag{4.5}$$

(2) Integrate both sides of the equation over the specified interval $t_2 - t_1$ (4.6).

$$\int_{t_1}^{t_2} \Phi_j(t)x(t)dt = \int_{t_1}^{t_2} \Phi_j(t)\left[\sum_{n=0}^{N} a_n \Phi_n(t)\right]dt = \sum_{n=0}^{N} a_n \int_{t_1}^{t_2} \Phi_j(t)\Phi_n(t)dt \tag{4.6}$$

Applying the condition of orthogonality, the right side of the equation becomes (4.7).

$$\int_{t_1}^{t_2} \Phi_j(t)x(t)dt = a_j\lambda_j \tag{4.7}$$

Note that when the indexes of the basis function are equal, $j = n$, the resulting integral will be some value, lambda (4.8);

$$\int_{t_1}^{t_2} \Phi_j(t)\Phi_n(t)dt = \lambda_j \tag{4.8}$$

otherwise, if indexes of the basis function are not equal, $j \neq n$, the integral will equal zero. The equation may be rewritten to solve for the coefficients as shown in the following equation:

$$\frac{1}{\lambda_j}\int_{t_1}^{t_2} \Phi_j(t)x(t)dt = a_j \tag{4.9}$$

Recall the general equation for an energy or average power signal as shown in the following equation:

$$E = \int_{t_1}^{t_2} x^2(t)dt \tag{4.10}$$

By substituting the general basis function representation of a signal (4.11) into (4.10);

$$x(t) = \sum_{n=0}^{N} a_n\Phi_n(t) \tag{4.11}$$

the energy equation (4.10), may be expressed as (4.12):

$$E = \int_{t_1}^{t_2} x^2(t)dt = \int_{t_1}^{t_2} x(t)\sum_{n=0}^{N} a_n\Phi_n(t)dt = \sum_{n=0}^{N} a_n\int_{t_1}^{t_2} \Phi_n(t)x(t)dt \tag{4.12}$$

From the orthogonality condition, (4.12) is rewritten in terms of the coefficient and lambda as given in (4.13),

$$E = \sum_{n=0}^{N} a_n(a_n \lambda_n) \tag{4.13}$$

or as given in (4.14).

$$E = \int_{t_1}^{t_2} x^2(t)dt = \sum_{n=0}^{N} a_n^2 \lambda_n \tag{4.14}$$

Since $a_n^2 \lambda_n$ is the energy in the nth basis function, each term of the summation is the energy associated with its index, nth component of the representation. Thus, the TOTAL energy of a signal is the sum of the energies of its individual orthogonal coefficients, which is often referred to as "Parseval's theorem."

Many functions are, or can be, made orthogonal over an interval, but it does not mean that the function may be a desirable function to use as a basis function. As engineers, we tend to use the trigonometric functions, sine and cosine, in many analytical applications. Why are the sinusoids so popular? Three important facts about sinusoidal expressions stand out:

(1) Sinusoidal functions are very useful because they remain sinusoidal after various mathematical operations (i.e., integration, derivative). Also, exponents of sinusoidal functions can be expressed as exponents by Euler's identity $(\cos \theta + j \sin \theta = e^{j\theta})$.

(2) The sum or difference of two sinusoids of the *same frequency* remains a sinusoid.

(3) This property combined with the superposition properties of linear systems implies that representing a signal as a sum of sinusoids may be a very convenient technique, which is used for periodic signals.

4.4   SIGNAL REPRESENTATIONS

4.4.1   Fourier Transforms

The Trigonometric Fourier Transform Function. As indicated in the previous section, it is convenient to use basis functions that are invariant with respect to the mathematical operations, that is, integration, derivative, or summation. The "Sinusoidal Basis Function" has these properties and is commonly used to represent periodic, complex signals. It is possible to use sines and cosines as basis functions as given by (4.15), but it is more convenient to use complex exponentials and to write the "Exponential Fourier Transform" as (4.16).

$$x(t) = a_0 + \sum_{n=1}^{\infty} [a_n \cos(n\omega_0 t) + b_n \sin(n\omega_0 t)] \tag{4.15}$$

$$x(t) = c_n e^{-jn\omega_0 t} \tag{4.16}$$

You should be familiar with the relationship given in (4.17), and note that the summation may be positive or negative.

$$e^{\pm jn w_0 t} = \cos(n w_0 t) \pm j \sin(n w_0 t) \tag{4.17}$$

Consider a function of time for which you would like to obtain the Fourier series representation. The function is periodic over the interval, $T = t_2 - t_1$. For the series to converge to the true value, the Dirichlet conditions requires that the function

1.   be single-valued within the interval, $t_2 - t_1$;

2.   have a finite number of maximums and minimums in a finite time;

3.   satisfy the inequalities of (4.18)

$$\int_{t_1}^{t_2} |x(t)| \, dt < \infty; \text{ and} \tag{4.18}$$

4.   be a finite number of discontinuities.

For the exponential Fourier Series, the basis function is defined as given by (4.19).

$$e^{\pm jnw_0t} = \Phi_n(t), \text{ where } n = \pm(0, 1, 2, \ldots., \infty), \text{ and} \tag{4.19}$$

where: $\omega_0 \frac{2\pi}{T}$ (the fundamental frequency is in radians).

To prove that the functions are orthogonal over the interval start with (4.20).

$$\int_{t_1}^{t_1+T} e^{jnw_0t} * e^{-jnw_0t}dt = 0 \quad \text{when} \quad n \neq k$$
$$= T \quad \text{when} \quad n = k \tag{4.20}$$

The coefficients for the series can be expressed as (4.21).

$$a_n = \frac{1}{T} \int_{t_1}^{t_1+T} x(t)e^{-jnw_0t}dt \tag{4.21}$$

The coefficients are usually complex and can be expressed as $a_{-n} = \alpha_n$ * (* means conjugate). The signal $x(t)$ is expressed as (4.22):

$$x(t) = \sum_{n=-\infty}^{\infty} \alpha_n e^{jnw_0t} \tag{4.22}$$

The accuracy of an approximate representation using a finite number of terms is obtained from the energy ratio given by (4.23) and (4.24).

$$\eta_m = \frac{\text{error energy}}{\text{signal energy}} = 1 - \frac{1}{E}\sum_{n=0}^{M} \lambda_n a_n^2 \tag{4.23}$$

where

$$E = \int_{t_1}^{t_2} x^2(t)dt \tag{4.24}$$

For the Fourier series, $\lambda_n = T$, since $\alpha_{-n} = \alpha_n^*$. Then, $|\alpha_n|^2 = |\alpha_{-n}|^2$.

Then from $-M$ to M, the fractional error is given by (4.25):

$$\eta_m = 1 - \frac{1}{E} \sum_{n=0}^{M} T |\alpha_n|^2$$

$$\eta_m = 1 - \frac{1}{E} \left[\alpha_0^2 + 2 \sum_{n=0}^{M} T |\alpha_n|^2 \right]$$

(4.25)

Example: Consider the periodic sequence, $x(t) = A$ for $0 < t < t_a$, and $x(t) = 0$ for $t_a < t < T$: as in (4.26) and (4.27).

$$\alpha_n = \frac{1}{T} \int_0^{t_a} A e^{-jnw_0 t} dt + 0 \text{ (for } t_a < t < T)$$

(4.26)

$$\alpha_n = \frac{A}{T} \frac{-1}{jnw_0} e^{-jnw_0 t} \Big|_0^{t_a}$$

$$\alpha_n = \frac{A}{T} \frac{-1}{jnw_0} e^{-jnw_0 t_a} - \frac{A}{T} \left(\frac{-1}{jnw_0} \right) (1)$$

(4.27)

$$\alpha_n = \frac{A}{jnw_0 T} \left(1 - e^{-jnw_0 t} \right)$$

By rewriting the terms in the bracket, you get (4.28):

$$e^{-j\frac{x}{2}} \left(e^{j\frac{x}{2}} - e^{-j\frac{x}{2}} \right), \text{ where } x = nw_0 t$$

(4.28)

and α_n becomes (4.29):

$$\alpha_n = \left[\frac{e^{j\frac{nw_0 t_a}{2}} - e^{-j\frac{nw_0 t_a}{2}}}{jnw_0} \right] \left(e^{-j\frac{nw_0 t_a}{2}} \right)$$

(4.29)

The [ ] (square bracket) term is a trig identity [(4.30) and (4.31)] with circular functions where

$$\sin(x) = \frac{e^{jx} - e^{-jx}}{2j}$$

(4.30)

$$\text{With } x = \frac{nw_0 t_a}{2}$$

(4.31)

So, to begin solving this problem, you first need to multiply the denominator by 2/2, as shown by (4.32), (4.33), and (4.34).

$$\alpha_n = \frac{A}{T}\frac{2}{2}\frac{\sin\left(nw_0\frac{t_a}{2}\right)}{nw_0/2}e^{-j\frac{nw_0 t_a}{2}} \tag{4.32}$$

Then, multiply by t_a/t_a.

$$\alpha_n = \frac{A}{T}\frac{t_a}{t_a}\frac{\sin\left(nw_0\frac{t_a}{2}\right)}{\frac{nw_0}{2}}e^{-j\frac{nw_0 t_a}{2}} \tag{4.33}$$

$$\alpha_n = \frac{At_a}{T}\left[\frac{\sin\left(nw_0\frac{t_a}{2}\right)}{\frac{nw_0 t_a}{2}}\right]e^{-j\frac{nw_0 t_a}{2}} \tag{4.34}$$

Since $w_0 = 2\pi f_0$ and $f_0 = 1/T$, then $w_0 = (2\pi)/T$. Thus, alpha can be rewritten as (4.35).

$$\alpha_n = \frac{At_a}{T}\left[\frac{\sin(\pi nt_a/T)}{n\pi t_a/T}\right]e^{-j\frac{2n\pi t_a}{2T}} \tag{4.35}$$

The complete series is written as (4.36) or equations in (4.37).

$$x(t) = \sum_{n=-\infty}^{\infty}\alpha_n e^{jnw_0 t} \quad \text{or} \tag{4.36}$$

$$x(t) = \sum_{n=-\infty}^{\infty}\left\{\frac{At_a}{T}\left[\frac{\sin(n\pi t_a/T)}{\frac{n\pi t_a}{T}}\right]e^{-j\frac{2n\pi t_a}{2T}}e^{j\frac{n2\pi t}{T}}\right\}$$

$$x(t) = \sum_{n=-\infty}^{\infty}\left\{\frac{At_a}{T}\left[\frac{\sin(n\pi t_a/T)}{n\pi t_a/T}\right]e^{-j\frac{2n\pi\left(t-\frac{t_a}{2}\right)}{T}}\right\} \tag{4.37}$$

Keep in mind that the signal is being represented in terms of sinusoids having frequencies that are multiples of the fundamental frequency $1/T$. The coefficients, α_n, give the magnitude and phase of these sinusoids, which constitute an approximate frequency-domain description of the explicit time-domain signal. The Fourier series is by far the most commonly used orthogonal representation.

4.4.2 Legendre Functions

Other signal representation includes the Legendre functions, which are an orthonormal set of basis functions in the time interval from -1 to 1. The general basis function is given by (4.38) and (4.39).

$$\Phi_n(t) = \sqrt{\frac{2n+1}{2}} P_n(t) \quad \text{for} \quad -1 \leq t \leq 1. \tag{4.38}$$

where $P_n(t)$ is the Legendre polynomials, given by (4.39) and (4.40).

$$P_n(t) = \frac{1}{2^n n!} \frac{d^n}{dt^n} (t^2 - 1)^n \quad \text{for} \quad n = 0, 1, 2, \ldots \text{ for } -1 \leq t \leq 1 \tag{4.39}$$

$$\Phi_0(t) = \frac{1}{\sqrt{2}}$$

$$\Phi_1(t) = \sqrt{\frac{3}{2}} t$$

$$\Phi_2(t) = \sqrt{\frac{5}{2}} \left(\frac{3}{2} t^2 - \frac{1}{2} \right)$$

$$\Phi_3(t) = \sqrt{\frac{7}{2}} \left(\frac{5}{2} t^3 - \frac{3}{2} t \right)$$

$$\tag{4.40}$$

These basis functions may be convenient when the signals have a predominant linear or quadratic term.

4.4.3 Laguerre Functions

When the time interval of the representation is from 0 to ∞, the Laguerre functions form a complete orthonormal set defined by (4.41):

$$\Phi_n(t) = \frac{1}{n!} e^{-\frac{t}{2}} \ln(t) \quad \text{for} \quad 0 \leq t < \infty \tag{4.41}$$

ln is not natural logarithm, but rather as defined by (4.42):

$$\ln(t) = e^t \frac{d^n}{dt^n} (t^n e^{-t}) \tag{4.42}$$

4.4.4   Hermite Functions

Hermite functions (4.43) are orthonormal over the range from $-\infty$ to ∞.

$$\Phi_n(t) = (2^n n! \sqrt{\pi})^{-1/2} e^{-\frac{t^2}{2}} H_n(t)$$

$$H_n(t) = (-1)^n e^{t^2} \frac{d^n}{dt^n}(e^{-t^2}) \qquad (4.43)$$

Cardinal functions are given by (4.44).

$$\Phi_n(t) = \frac{\sin \pi (2wt - n)}{\pi (2wt - n)} \text{ for } n = 0, \pm 1, \pm 2, \ldots \qquad (4.44)$$

Another property of orthogonal functions has to do with the accuracy of the representation when not all of the terms can be used for an exact representation. In most cases, N is infinity for exact representation, but the coefficient, a_n, becomes smaller as n increases.

The approximation or estimation of $x(t)$ is expressed as $x'(t)$ as in (4.45).

$$x'(t) = \sum_{n=0}^{M} a'_n \Phi_n(t), \text{ where } M \text{ is finite.} \qquad (4.45)$$

A measure of the closeness or goodness of approximation is the integral squared error (4.46).

$$I = \int_{t_1}^{t_2} [x(t) - x'(t)]^2 dt \quad \text{and} \quad I = 0 \text{ when } x'(t) = x(t) \qquad (4.46)$$

The smaller the value of I, the better the approximation. The integral squared error is applicable *only* when $x(t)$ is an energy signal or when $x(t)$ is periodic. When $x(t)$ is an energy signal, the limits of integration may range from $-\infty$ to $+\infty$. When $x(t)$ is periodic, then $t_2 - t_1$ is equal to the period T.

To find the optimum values of the estimator for the basis coefficient, we use the integral squared error or sum of squared error (SSE) as shown in (4.47).

$$I = \int_{t_1}^{t_2} \left[x(t) - \hat{x}(t) \right]^2 dt \qquad (4.47)$$

Next substitute (4.48) into (4.47) to obtain the least squared error (I), which is given by (4.49).

$$\hat{x}(t) = \sum_{n=0}^{M} \hat{a}_n \phi_n(t) \tag{4.48}$$

$$I = \int_{t_1}^{t_2} \left[x(t) - \sum_{n=0}^{M} \hat{a}_n \phi_n(t) \right]^2 dt \tag{4.49}$$

The next step is to find the set of estimators coefficients $\hat{a}_n$ that minimize the value of the integral (least squared error) in (4.47). One method is to differentiate the least squared error (I) with respect to each of the coefficients and set each derivative equal to 0. There are other alternative procedures that may be used, but will not be discussed in this chapter.

CHAPTER 5

Data Acquisition Process

At this point, I would like to change pace and have a discussion of general steps or procedures in data acquisition and processing. The required operations may be divided into 5 primary categories (Fig. 5.1):

1. Data collection,

2. Data recording/transmission,

3. Data preparation,

4. Data qualification, and

5. Data analysis.

5.1   DATA COLLECTION

The primary element in data collection is the instrumentation transducer. In general, a transducer is any device which translates power from one form to another. I will not get into this but in general a model might be as shown in Fig. 5.2.

In some cases it is possible to perform all desired data processing directly on the transducer signals in real time. In most cases, this is not practical. After preamplification some initial filtering and postamplification, most signals are recorded on magnetic media. At the present time, some form of storage of the amplified and preconditioned transducer signals are still required.

FIGURE 5.1: Data acquisition and analysis process

5.2 DATA RECORDING/TRANSMISSION

As in the past, magnetic storage systems are the most convenient and desirable type of data storage. In the 1950s through the 1980s, magnetic tape recordings were the only way to store large quantities of data. If a magnetic tape recorder was used, the most widely used method was referred to as "Frequency Modification" (FM) for analog type recordings. Beyond the limitation inherent in the magnetization-reproduce and modulation-demodulation operations, there are a number of miscellaneous problems.

Among the more important of these problems are the errors associated with variations in the speed at which the tape passes over the record and/or reproduce heads. These errors were called "Time Based Errors," with the most common being "Flutter," which is the variation of tape velocity from the normal. For the frequency modulation (FM) recording systems, the time based (frequency) is used to represent the amplitude variations of the signal. The error in good high-quality FM recorders is about 0.25%.

Currently, few laboratories still use magnetic tape recordings. The large "Giga-bite" disk memories and optical disc with compression techniques are the most common approach to storage of data; however, the data must be in digital format, and thus must be converted from an analog format to a digital format. The advantage of storing the raw data is that different signal-processing methods may be used on the same data and the results compared.

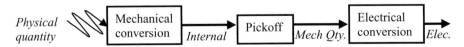

FIGURE 5.2: General model of a mechanical transducer for data collection, that is, a blood pressure transducer

5.3   DATA PREPARATION

The third step is data preparation of the *raw data* supplied as voltage time histories of some physical phenomenon. Preparation involves three distinct processes as shown in Fig. 5.3:

1. editing;

2. conversion; and

3. preprocessing.

Data editing refers to preanalysis procedures that are designed to detect and eliminate superfluous and/or degraded data signals. The undesirable parts of the signal may be from acquisition or recording problems that result in excessive noise, movement artifacts, signal dropout, etc. Editing may be through visual inspection of graphic recordings. Normally, experts in the field are invited to perform the inspection and identify those sections of the data that should not be analyzed. Some laboratories use a Hewlett-Packard Real-time spectrum analyzer and preview the data. In short, "Editing is critical in digital processing" and it is best performed by humans, and not by machines.

Data preparation beyond editing usually includes conversions of the physical phenomena into engineering units (calibration) and perhaps digitization. Digitization consists of the two separate and distinct operations in converting from analog to digital signals:

1. *Sampling interval*, which is the transition from continuous time to discrete time, and

2. *Sample quantization*: the transition from a continuous changing amplitude to a discrete amplitude representation.

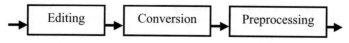

FIGURE 5.3: Data preparation

Considerations that must be taken into account when digitizing an analog signal will be covered in two chapters: one on Sampling Theory and the second on Analog-to-Digital Conversion.

Prepossessing may include formatting of the digital data (Binary, ASCII, etc., format), filtering to clean up the digital signal, removing any DC offset, obtaining the derivative of the signal, etc. But before getting into digital conversion, let us continue with the data-acquisition process by examining the fourth step of "Random Data Qualification."

5.4   RANDOM DATA QUALIFICATION

Qualification of data is an important and essential part in the statistical analysis of random data. Quantification involves three distinct processes as shown in Fig. 5.4:

1. stationarity,

2. periodicity, and

3. normality.

It is well known in statistics that researchers should not use "parametric statistical methods" unless the data are shown to be "Gaussian Distributed" and "Independent." If the data are not Gaussian Distributed and Independent, parametric statistics cannot be used and the researcher must then revert to "nonparametric statistical methods."

Likewise, necessary conditions for Power Spectral Estimation require that the random data be tested for "Stationarity" and "Independence," which may be tested parametrically by showing the mean and autocorrelation of the random data to be time invariant or by using the nonparametric "RUNS" Test.

After showing the data to be stationary, researchers may examine the data for "Periodicities" (cyclic variations with time). Periodicity may be analyzed via the

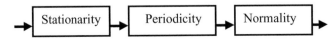

FIGURE 5.4: Data qualification

FIGURE 5.5: Random data analysis

"Correlation Function" using either Time or Frequency Domain approach. Periodicity may also be analyzed via the "Power Spectral Analysis." The next step regarding random data may be to examine the distribution of the data ("Is the data Normality distributed?") or to determine all the basic statistical properties (Moments) of the signal (data).

5.5   RANDOM DATA ANALYSIS

The fifth and final step is data analysis of the data supplied as time histories of some physical phenomenon. Data analysis involves three distinct and independent processes dependent on the experimental question and the data set as shown in Fig. 5.5:

1. analysis of individual records,

2. analysis of multiple records, and

3. linear relationships.

If the analysis is to be accomplished on an individual record, that is, one channel of electrocardiogram recording, or one channel of electroencephalogram recording, etc., then either "Autocorrelation Function Analysis" (Time-domain Analysis) or "Autospectral Analysis" (Frequency-domain Analysis) would be preformed on the random data.

If the analysis is to be accomplished on multiple records, that is, more than one channel of recording, then either "Cross-correlation Function Analysis" (Time-domain Analysis) or "Cross-spectral Analysis" (Frequency-domain Analysis) would be preformed on the random data. Multiple Records would be used to determine the "Transfer Functions" of systems when the input stimulus and the output responses are random signals.

In a relational study, cause and effect, the primary aim would be to determine the relationship of a stimulus to the response. The most popular method for determining linear relationships is to use the least squares regression analysis and to determine the correlation coefficient; however, the least squares regression analysis is difficult to apply in many random data cases. Therefore, an alternative method is to determine the model or transfer function via auto- and cross-spectra. In addition, for any model or transfer function, it is necessary to determine how well the transfer functions represent the system process; hence, the investigator must perform a "Coherence Function Analysis" to determine the "Goodness Fit."

C H A P T E R   6

Sampling Theory and Analog-to-Digital Conversion

6.1   BASIC CONCEPTS

Use of a digital computer for processing an analog signal implies that the signal must be adequately represented in digital format (a sequence of numbers). It is therefore necessary to convert the analog signal into a suitable digital format.

By definition, an **analog signal** is a continuous physical phenomenon or independent variable that varies with time. Transducers convert physical phenomena to electrical voltages or currents that follow the variations (changes) of the physical phenomena with time. One might say that an electrical signal is an *Analog Signal* if it is a continuous voltage or current that represents of a physical variable. Analog signals are represented as a deterministic signal, $x(t)$, for use in a differential equation. Or, one may simply state that an analog signal is any signal that is not a digital signal, but then you must define a digital signal. There are three classes of analog signals.

a)   Continuous in both time and amplitude (Fig. 6.1).

b)   Continuous in time (Fig. 6.2) but taking on only discrete amplitude values, that is, analog signals used to approximate digital bit streams in serial communication, which are used primarily for time codes or markers.

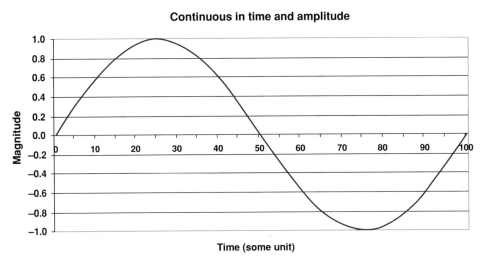

FIGURE 6.1: Analog signal that is continuous in both time and magnitude

c) Continuous in amplitude (Fig. 6.3) but defined only at discrete points in time, for example, sampled analog signals.

By definition, a digital signal is an ordered sequence of numbers, each of which is represented by a finite sequence of bits; that is, of finite length. It is important to note that a digital signal is defined only at discrete points in time and takes on only one of a

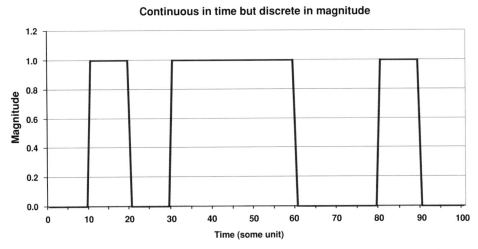

FIGURE 6.2: Analog signal that is continuous in both time and discrete in magnitude

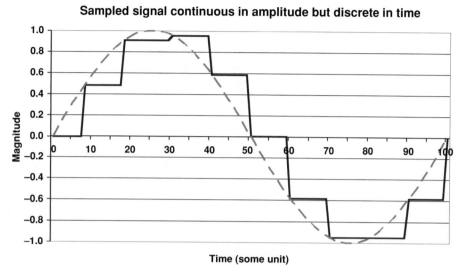

FIGURE 6.3: An analog signal that has been sampled such that the sampled signal is "continuous in amplitude but defined only at discrete points in time." The dashed line is the original sinusoidal signal

finite set of discrete values at each of these points. A digital signal can represent samples of an analog signal at discrete points in time.

There are two distinct operations in converting from analog to digital signals:

1. *Sampling*, which is the transition from continuous-time to discrete-time

2. *Sample quantization*, the transition from a continuous changing amplitude to a discrete amplitude representation

Reversing the process from digital to analog involves a similar two-step procedure in reverse.

6.2   SAMPLING THEORY

Before getting into the topic on analog-to-digital conversion, it is necessary to have some knowledge and understanding of the theory behind the sampling of data. The mechanism for obtaining a sample from an analog signal can be modeled in two ways. Ideally, an instantaneous value of the signal is obtained at the sample time. This model

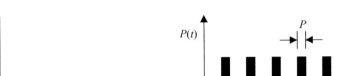

FIGURE 6.4: The realistic Delta function model. Note the uniform magnitude and finite pulse width

is known as a "delta function." The model is attractive in analytical expressions, but the delta function (model) is really impossible to generate in the real world with real circuit components. The Realist model is considered as modulation of the original signal by a uniform train of pulses, with a pulse finite width of "P" and interval of the sample period "T," as shown in Fig. 6.4.

Each pulse represents the instant of time that the value of an analog signal is multiplied by the delta function. It may be easier to think of the process as grabbing a value from the analog signal at each pulse such that the resulting collection of values is no longer continuous in time but only exists at discrete points in time.

Sampling Theorem: The sampling theory is generally associated with the works of Nyquist and Shannon. The Sampling Theorem for band-limited signals of finite energy can be interpreted in two ways:

1. Every signal of finite energy and Bandwidth W Hz may be completely recovered from the knowledge of its samples taken at the rate of 2 W per second (called the Nyquist rate). The recovery is stable in the sense that a small error in reading sample values produces only a corresponding small error in the recovered signal.

2. Every square-summable sequence of numbers may be transmitted at the rate of 2 W per second over an ideal channel of bandwidth W Hz. By being represented as the samples of a constructed band limited signal of finite length.

What does it all mean? It has been proven by Landau that

(1) stable sampling can*not* be performed at a rate lower than the Nyquist rate; and

(2) data can*not* be transmitted as samples at a rate higher than the Nyquist rate.

To put it in simpler terms, Lessard's paraphrase of the sampling theorem is: "Sampling rate must be at least twice the highest frequency of interest in the signal."

The Fourier transform of the periodic analog signal, $x(t)$, is given by (6.1):

$$\tilde{f}(x(t)) = X(j\omega) = \int_{-\infty}^{\infty} x(t)e^{-j\omega t}dt \tag{6.1}$$

When the analog signal is sampled, you have the resulting *sampled* signal (6.2):

$$x_s(t) = x(t)p(t) \tag{6.2}$$

where $p(t)$ is the sampling function with the period T.

The Fourier Series of the sampling function is as shown in (6.3):

$$p(t) = \sum_{n=-\infty}^{\infty} c_n e^{j w_s nt} \tag{6.3}$$

where $w_s = 2\pi/T$ (sampling rate in radians/second) and c_n is the Complex Fourier coefficient.

The Fourier coefficients are evaluated by (6.4):

$$c_n = \frac{1}{T} \int_{-T/2}^{T/2} p(t)e^{-j\omega nt}dt \tag{6.4}$$

The result is given by (6.5):

$$c_n = \frac{P}{T}\left(\frac{\sin\left(n\omega_s \, {}^{P}/_{2}\right)}{n\omega_s \, {}^{P}/_{2}}\right) \tag{6.5}$$

For the Sampling Function, where P is the pulse width of $p(t)$, the Fourier Transform of the Sampled function (6.6) is written in the integral format as (6.7),

$$x_s(t) = x(t)p(t) = \sum_{n=-\infty}^{\infty} x(t)c_n e^{j w_s nt} \tag{6.6}$$

$$\tilde{f}(x_s(t)) = \int_{-\infty}^{\infty} \left[\sum_{n=-\infty}^{\infty} x(t)c_n e^{j w_s nt} \right] e^{-j\omega t} dt \tag{6.7}$$

which results in the expression given by (6.8):

$$\sum_{n=-\infty}^{\infty} c_n X(\omega - n\omega_s) \tag{6.8}$$

where for the ideal delta function, $c_n = 1/T$ and $t = nT$.

The Fourier Transform of the sampled function is often called "Discrete Fourier Transform of a Time Series." It should be noted that the Discrete Fourier Transform DFT is a transform in its own right as is the Fourier integral transform, and that the DFT has mathematical properties analogous to the Fourier Integral Transform. The usefulness of the DFT is in obtaining the Power Spectrum analysis or filter simulation on digital computers. The Fast Fourier Transform, which will be covered, is a highly efficient procedure of a time series. Restrictions on the period T of the pulse train $p(t)$ to avoid ambiguity in the sampled version. The Nyquist Sampling Theorem must be followed.

Let us examine the 1st step in converting from an analog signal to a digital signal. Consider an analog signal $v(t) = V \sin(\omega t)$. The sampled values will give the exact value of $v(t)$ only at the sampled time; thus, a discrete-time sample of $v(t)$. If the period or time between samples is constant, T (the sampled period), the discrete approximation is expressed as $v(nt)$, where n is an integer index. Sampling with a constant period is referred to as "uniform sampling," which is the most widely used sampling method because of the ease of implementing and modeling. As is known, the sample period, T, effects how accurately the original signal is represented by the sampled approximation. The ideal sampled signal $X_s(t)$ is zero except at time $t = nT$ and the signal is represented

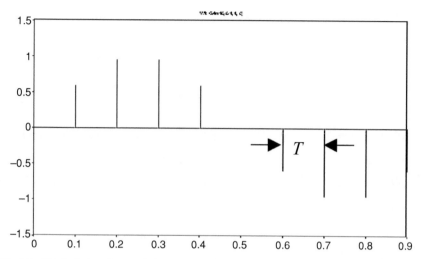

FIGURE 6.5: Digital signal sampled at a rate of $1/T$, where T is the sampling interval

as a sequence of numbers $x(n)$ representing sampled values. Figure 6.5 shows a sampled sinusoid with a sampling interval of 1/10 the period of the sinusoid.

$$v(t) = \sin(t) \qquad T = 2\pi/10$$

6.3   QUANTIZATION

Now, let us consider the quantization operation since the magnitude of each data sample must be expressed by some fixed number of digits, since only a fixed set of levels is available for approximating the infinite number of levels in continuous data. The accuracy of the approximating process is a function of the number of available levels. Since most digitizers produce binary outputs compatible with computers, the number of levels may be described by the number of binary digits (bits) produced.

6.3.1   Resolution or Quantization Step

In digital systems, sampled values must be represented in some binary number code that is stored in finite-length register. A binary sequence of length n can represent at most 2^n different numbers: that is, a 4-bit register can represent 2^4 or 16 numbers from 0 to 15 inclusively. If the amplitude of the analog signal is to be between $-v$ and $+v$ volts, the interval can be divided at most into 2^n discrete numbers. This interval is divided into quantization step size of

$$\frac{2(v)}{2^n} = E_0 \tag{6.9}$$

The size of the quantization step (E_0) or aperture introduces an amplitude approximation which limits the potential accuracy of the subsequent processing. For example, a 4-bit A/D converter with an input range of -5 to $+5$ volts had a resolution (E_0) of 10 volts divided by 16 levels, results in a quantizing step with a voltage resolution of 0.625 volts as given by (6.10):

$$\frac{10v}{2^4} = \frac{10}{16} = 0.625 \text{ volts} \tag{6.10}$$

A change in the least significant bit will cause a step from one level to the next. The most common types of sample quantization are as follows:

1) *Rounding*: The quantization level nearest the actual sample values is used to approximate the sample;

2) *Truncation*: The sample value is approximated by the highest quantization level that is *not* greater than the actual value;

and

3) *Signed Magnitude Truncation*: Positive samples are truncated but negative values are approximated by the nearest quantization level that is not less than the actual sample values.

The digital signal $X(nT)$ can often be expressed as two components: $X(nT) = X_0(nT) + e(nT)$, where $X_0(nT)$ can be thought of as the actual signal value at $t = nT$ and $e(nT)$ is an added error signal called "Quantization noise." The error signal $e(nT)$ is modeled as a uniformly distributed random sequence with the exact nature of the distribution dependent on the type of quantization involved. The variance for Truncation and rounding is the same as shown in (6.11):

$$\sigma_e^2 = \frac{E_0^2}{12} \qquad (6.11)$$

where E_0 is the quantization step size.

6.3.2 Signal-to-Noise Ratio of an A/D Converter

The signal-to-noise ratio introduced by the quantization is defined as the ratio of peak-to-peak signal to RMS noise in decibels (6.12).

$$S/N = 20 \log \left(2^n \sqrt{12} \right) \mathrm{dB} \cong 6.02n + 10 \qquad (6.12)$$

Hence one should know that the signal-to-noise ratio increases by approximately 6 dB for each added bit in the converter.

6.3.3 Dynamic Range of an A/D Converter

The dynamic range of the converter is defined as the ratio of the full-scale voltage to quantization step size as given by (6.13):

$$\text{Dynamic range } = 20 \ \log 2^n = 6.02n \ \mathrm{dB} \qquad (6.13)$$

The required dynamic range of an analog-to-digital converter (ADC) is an important attribute of a physical problem if the amplitude of the signal is to be retained for processing. One would think that increasing the number of bits (n), which decreases the quantization size, would lead to any arbitrary degree of accuracy. But recall that analog signals are limited in accuracy by the thermal noise of the analog components. Real ADC converters are analog devices, which exhibit other types of instrumentation errors, that is,

a) Offset errors

b) Gain error or scale factor

c) *Nonlinearity*: Linear offset and gain errors may be adjustable with a potentiometer; however, most nonlinearities tend to be uncorrectable.

Digital Codes: The next problem is to code the numbers from the ADC to a format suitable for further computation. So what are the most common formats used in digital processing of fixed points? If one thinks about it, 2^n different numbers can be represented by an n-bit digital word. The position of the binary point is assumed fixed. Bits to the left of the point represent the integer part. Bits to the right of the point represent the fractional part.

There are three common fixed-point formats for representing negative numbers.

1. *Sign magnitude*: The left most bit (most significant bit) represents the sign of the number, and the remaining bits $(n - 1)$ represent the magnitude of the number. A well-known problem is that in a 4-bit sign magnitude, both the codes 1000 and 0000 represent the value of 0.

2. *One's compliment*: Positive numbers are represented in sign magnitude format. The negative of a number is obtained by complimenting all the bits of the number, the negative of 0101 is 1010. But again, there are two representations for 0, which are 0000 and its compliment 1111.

3. *Two's compliment* : The representation depends on the sign of the number.
   a) Positive numbers are represented in sign-magnitude format.
   b) Negative numbers are represented in two's compliment.

To obtain the two's compliment requires two operations; first, take the one's compliments, and then add 1 to the least significant (right most) bit. For example, The negative of $0101 = 1011$. Zero (0) is then 0000 (only one representation), with the largest positive number being 0111 for $n = 4$, while the largest negative number is 1000.

6.4   DIGITAL FORMAT

6.4.1   Floating Point Format

Floating point representation of a number consists of two fixed-point numbers representing a mantissa and an exponent. The floating-point number, n, is the product of mantissa, m, and raising base (2) to exponent a, as shown in (6.14).

$$n = m2^a \qquad (6.14)$$

Sign of a floating-point number is represented by the sign bit of the mantissa while the sign bit of the exponent indicates a negative for numbers with magnitudes less than 0.5. The mantissa is normalized to lie within the range, as shown in (6.15).

$$\frac{1}{2} \leq m < 1 \qquad (6.15)$$

6.4.2   Block Floating Point Format

Another format is the block floating point format where a single exponent is used for a block of fixed points. Most ADC produce fixed point numbers. Some ADC contain analog prescaling that allows sample values to be output in floating point or block floating point. Some ADC encode as differential encoding. The output is not the absolute value of the sample but rather the difference between successive sample values can be represented with fewer bits.

6.5   SIGNAL RECONSTRUCTION—DIGITAL-TO-ANALOG CONVERSION (DAC)

Two steps, which are the reverse of Analog-to-Digital Conversion (ADC), are required to reconstruct the analog signal from digitized data. The first step is to transform from discrete amplitude representation to continuous value. Second, transform the discrete time sequence to a continuous-time sample.

In most cases, a sample and hold circuit is used to hold the sample value constant until the next sample value. The results would look like the solid trace in Fig. 6.3. Usually, the zero-order hold circuit is followed by a low pass filter to remove the high-frequency components from sharp amplitude charges.

CHAPTER 7

Stationarity and Ergodic Random Processes

7.1 INTRODUCTION

In Chapter 3, signals were classified as either deterministic or random. Random signals can be further subdivided and classified as stationary and nonstationary. Properties necessary to a stationary process were provided for first-order, second-order, wide-sense, and strictly stationary. Two different methods of determining the functions necessary for a random process to be classified as a "Stationary Process," which include ensemble and extension in time, are discussed in this chapter. In addition, the necessary conditions for a random process to be an "Ergodic Process" are also presented. Nonparametric tests and examples are also included to provide the reader with the necessary information and tools to identify or classify random processes as being a "Stationary Process" and/or an "Ergodic Process."

In qualifying data, one must first determine whether the signal is deterministic or random. If a signal can be predicted exactly for a particular time span, it is said to be deterministic. Examples of deterministic signals are given in (7.1), (7.2), and (7.3).

$$x(t) = 10 \sin 2\pi t \qquad \text{Sine wave} \qquad (7.1)$$
$$x(t) = 1, t > 0, \text{ and} \qquad \text{Unit step} \qquad (7.2)$$
$$x(t) = 0, t \geq 0$$
$$x(t) = 1 - e^{-t}, t \geq 0, \text{ and} \qquad \text{Exponential Response} \qquad (7.3)$$
$$x(t) = 0, t < 0$$

By definition, a random signal is an unpredictable signal; therefore, there is always some element of chance associated with it. Examples of random signals are given in (7.4), (7.5), and (7.6).

$$x(t) = 10 \ \sin \ (2\pi t + \theta) \tag{7.4}$$

where θ is a random variable uniformly distributed between 0 and 2π.

$$x(t) = A \ \sin \ (2\pi t + \theta) \tag{7.5}$$

where θ and A are independent random variables with known distributions.

$$x(t) = A \tag{7.6}$$

where A is a noise-like signal with no particular deterministic structure.

Random signals are further divided into those signals that are classified as "Stationary" and those that are classified as "Nonstationary." Only stationary random processes will be discussed, since available standard statistical analysis techniques are applicable only to stationary random processes.

A random process is "Stationary" if its statistical properties (mean and autocorrelation) are "Time Invariant." The two methods used in obtaining the statistical properties of any random process include the following:

1. *Ensemble method* : Where the assumption is that a large number of systems equipped with identical instruments are available and that a set of instantaneous values from multiple channels are obtained at the same time.

2. *Extension-in-time method (Ergodic Testing)* : Reduces the number of observations on a system to a single long record taken over an extended period of time.

7.2 ENSEMBLE METHOD TO TEST FOR STATIONARITY

Parameters derived from the Ensemble method will be expressed as $x(t)$. As seen in Fig. 7.1, parameters derived from the Ensemble method, such as the ensemble mean,

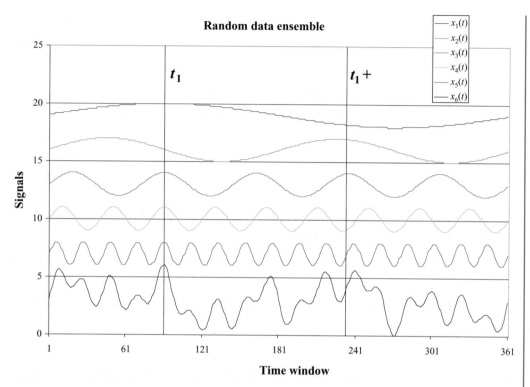

FIGURE 7.1: Random data ensemble. Values at an instant of time are taken across channels

require the selection of a specific time (t_1). Each function's value at t would be summed and divided by N (where N is the number of functions) to determine the ensemble average, where the subscript of x denotes a separate channel or function as in (7.7).

$$\overline{x}\,(t_1) = \frac{[x_1\,(t_1) + x_2\,(t_1) + x_3\,(t_1) + \cdots\cdots + x_N\,(t_1)]}{N} \tag{7.7}$$

Accordingly, a random process is stationary if the ensemble's averages are independent of time (referred to as *time invariant*) at which the group averages are computed, and if time averages are independent of functions making up this group. Thus, from the ensemble point of view, a stationary random process is one in which the statistics derived from observations on the ensemble members at any two distinct instants of time are the same (Fig. 7.1).

Most authors require further restraints for a process to be stationary. Some define several types of stationarity. It should be noted at this point that if a process can be proven to be stationary in even the weakest sense, stationarity of the random process can be assumed and analysis techniques requiring stationarity can then be conducted. Various types of stationarity are described in the subsequent section.

7.2.1 Statistics and Stationarity

7.2.1.1 First-Order Stationary Random Process

In statistics, the general method is to obtain a measure of a distribution by calculating its moments, that is, the 1st moment about the origin is often referred to as the "Mean" or "Average Value" about the origin, which is a measure of central tendency or centroid. The second moment is generally the mean about some point other than the origin, and the third moment is the variance of the distribution, which is a measure of disbursement.

If first-order statistics are independent of time (time invariant), it is implied that the moments of the ensemble values do not depend on time. The function associated with the random variable describing the values taken on by the different ensemble members at time t_1 is identical to the function associated with the random variable at any other time t_2.

First-order statistics for mean, mean square value, and variance are shown in (7.8), (7.9), and (7.10), respectively.

$$\overline{x}(t) = \int_{-\infty}^{\infty} x f_x(t) dt = \bar{x} \qquad\qquad \text{Mean Value} \qquad\qquad (7.8)$$

$$x^2(t) = \int_{-\infty}^{\infty} x^2 f_x(t) dt = x^2 \qquad\qquad \text{Mean Squared Value} \qquad (7.9)$$

$$T^2 x(t) = \int_{-\infty}^{\infty} [x - \bar{x}(t)]^2 f_x(t) dt = \sigma^2 \qquad \text{Variance} \qquad\qquad (7.10)$$

7.2.1.2 Second-Order Stationary Random Process

If all second-order statistics associated with random variables $X(t_1)$ and $X(t_2)$ are dependent at most on the time differences defined as $\tau = |t_1 - t_2|$ and not on the two values,

t_1 and t_2, the implications are that the joint function is obtained for any two random variables that describe the process at separate times. If a second-order random process is stationary, the first-order is also stationary. Second-order statistics are described by (7.11) and (7.12).

$$R_{xx}(\tau) = \int x(t)x(t+\tau)d\tau \qquad \text{Autocorelation function} \qquad (7.11)$$

$$L_{xx} = [x(t) - \bar{x}][[x(t-\tau) - \bar{x}] \qquad \text{Covariance} \qquad (7.12)$$

7.2.1.3 Wide-Sense Stationary

A random process is stationary in the wide sense if the mean from the first-order statistics and the autocorrelation from the second-order statistics are time invariant. Therefore, a process considered to be of second-order stationarity is also stationary in the wide sense. However, wide-sense stationarity does not imply second-order stationarity since covariance is not a statistic included in wide-sense stationarity.

7.2.2 Ensemble Method

7.2.2.1 Weakly Stationary

By definition, an ensemble is said to be a Stationary Random process as *Weakly Stationary* or *Stationary in the Wide Sense*, when the following two conditions are met:

1. **The mean,** $\mu_x(t)_1$, or the first moment of the random processes at all times, t, and

2. **The autocorrelation function,** R_{xx} $(t_1, t_1 + \lambda)$, or joint moment between the values of the random process at two different times do not vary as time t_1 varies.

The conditions are shown in (7.13) and (7.14).

$$\mu_x(t_1) = \mu_x = \lim_{N \to \infty} \frac{1}{N} \sum_{k=1}^{N} x_k(t_1) \qquad (7.13)$$

$$R_{xx}(t_1, t_1 + \lambda) = R_{xx}(\lambda) = \lim_{N \to \infty} \frac{1}{N} \sum_{k=1}^{N} x_k(t_1)x_k(t_1 + \lambda) \qquad (7.14)$$

7.2.2.2 Strictly Stationary

If one could collect all the higher moments and all the joint moments, then one would have a complete family of probability distribution functions describing the process. In the case where all possible moments and joint moments are time invariant, the random process is *Strongly Stationary* or *Strictly Stationary*. If a random process is strictly stationary, it must be stationary for any lower case.

7.3   EXTENSION-IN-TIME METHOD

The second method to test for stationarity, which deals with those random processes derived from the "Extension-in-Time Method." The extension-in-time method requires dividing a single long signal into segments of equal length of time, then the mean squared value (magnitude) of each segment is obtained. If the signal is a continuous analog signal, the average magnitude for the segment is given by (7.15).

$$\overline{X_1}(t) = \int_0^{t_1} x(t)dt \tag{7.15}$$

If the signal were a discrete sampled signal, then the segment average value would be obtained by dividing the sum of the data points in a segment by the total number of data points in the segment. From the extension-in-time point of view, a stationary random process is one in which the statistics measured from observations made on one system over an extended period of time are independent of the time in which observed time interval is located.

7.3.1   Ergodic Random Process

The previous section discussed how the properties of a random process can be determined by computing ensemble averages at specific instances of time. Often, it is possible to describe the properties of a stationary random process by computing time averages over specific sample functions in the ensemble. However, for the case of the extension-in-time method, the random process is referred to as an *Ergodic* process and the test is for *Ergodicity*.

Some processes that are stationary may also be categorized as Ergodic, but not "all" stationary processes are Ergodic. Yet, "all" Ergodic processes are considered weakly stationary. Orders of ergodicity can also be established. First-order ergodicity requires that all first-order statistics have equal ensemble and time series values. The same is true for second-order and weak-sense ergodicity. For most analyses techniques, weak-sense ergodicity is sufficient.

It should be noted that only Stationary Random Processes can be classified as either being *Ergodic* or being *Nonergodic*. In addition, if the signal (random process) is Ergodic, then the signal is also considered to be *Stationary in the Weak Sense*. It is also important to note that in accordance to Bendat and Piersol, that for many applications the verification of *Weak Stationarity* justifies the assumption of *Strong Stationarity*.

For example, one can obtain the mean value $\mu_x(k)$ and the autocorrelation function $R_x(\lambda, k)$ of the kth sample function with (7.16) and (7.17).

$$\mu_x(k) = \lim_{T \to \infty} \frac{1}{T} \int_0^T x_k(t)dt \tag{7.16}$$

$$R_x(\lambda, k) = \lim_{T \to \infty} \frac{1}{T} \int_0^T x_k(t)x_k(t + \lambda)dt \tag{7.17}$$

For the random process $\{x(t)\}$ to be stationary, the time averaged mean value $\mu_x(k)$, and the autocorrelation function $R_x(\lambda, k)$ of the segments should NOT differ (statistically) when computed over different sample functions; in that case, the random process is considered to be Ergodic.

7.4   REVIEW OF BASIC PARAMETRIC STATISTICS AND MOMENT-GENERATING FUNCTIONS

7.4.1   Moment-Generating Function

Given a random variable for continuous distribution as in (7.18),

$$M(t) \equiv \langle e^{tx} \rangle = \int e^{tx} P(x)dx \tag{7.18}$$

or for discrete distributions as given by (7.19).

$$M(t) \equiv \langle e^{tx} \rangle = \sum e^{tx} P(x) \qquad (7.19)$$

where $M(t) \equiv \langle e^{tx} \rangle$ is the moment-generating function. Expansion of (7.18) yields (7.20), the general moment-generating function.

$$M(t) = \int_{-\infty}^{\infty} (1 + tx + \frac{1}{2!}t^2 x^2 + \frac{1}{3!}t^3 x^3 + \cdots) P(x) dx$$

$$= 1 + tm_1 + \frac{1}{2!}t^2 m_2 + \frac{1}{3!}t^3 m_3 + \cdots + \frac{1}{r!}t^r m_r \qquad (7.20)$$

where m_r is the rth moment about zero.

For independent variables x and y, the moment-generating function may be used to generate joint moments, as shown in (7.21).

$$M_{x+y}(t) = \langle e^{t(x+y)} \rangle = \langle e^{tx} \rangle \langle e^{ty} \rangle = M_x(t) M_y(t) \qquad (7.21)$$

If $M(t)$ is differentiable at zero, the rth moments about the origin are given by the following equations (7.22):

$$\begin{aligned}
M(t) &= \langle e^{tx} \rangle, \quad \text{which is } M(0) = 1 \\
M'(t) &= \langle x e^{tx} \rangle, \quad \text{which is } M'(0) = \langle x \rangle \\
M''(t) &= \langle x^2 e^{tx} \rangle, \quad \text{which is } M''(0) = \langle x^2 \rangle \\
M'''(t) &= \langle x^3 e^{tx} \rangle, \quad \text{which is } M'''(0) = \langle x^3 \rangle \\
M^r(t) &= \langle x^r e^{tx} \rangle, \quad \text{which is } M^r(0) = \langle x^r \rangle
\end{aligned} \qquad (7.22)$$

Therefore, the mean and variance of a distribution are given by equations in (7.23).

$$\begin{aligned}
\mu &\equiv \langle x \rangle = M'(0) \\
\sigma^2 &\equiv \langle x^2 \rangle - \langle x \rangle^2 = M''(0) - \left[M'(0) \right]^2.
\end{aligned} \qquad (7.23)$$

7.5   SUMMARY

The moments may be simply computed using the moment-generating function. The nth raw or first moment, that is, moment about zero; the origin (μ'_n) of a distribution $P(x)$

is defined as $\mu'_n = \langle x^n \rangle$; where for Continuous Distributions (7.24):

$$\langle f(x) \rangle = \int f(x) P(x) dx \qquad (7.24)$$

and for Discrete Distributions (7.25)

$$\langle f(x) \rangle = \sum f(x) P(x) \qquad (7.25)$$

μ'_n, the mean, is usually simply denoted as $\mu = \mu_1$.

　　If the moment is taken about a point a rather the origin, the moment is termed the second moment and is generated by (7.26).

$$\mu_n(a) = \langle (x - a)^n \rangle = \sum (x - a)^n P(n) \qquad (7.26)$$

The moments are most commonly taken about the mean (a point a), which are also referred to as "Central Moments" and are denoted as μ_n. The Central or second moment about the origin is defined by (7.27), where $n = 2$.

$$\mu_n \equiv \langle (x - \mu)^n \rangle = \int (x - \mu)^n P(x) dx \qquad (7.27)$$

It should be noted that the third moment ($n = 3$) about the origin may also be called the second moment about the mean, $\mu_2 = \sigma^2$, which is equal to the variance or disbursement of the distribution; and the square root of the variance is called the standard deviation.

7.6   SUGGESTED READING

Papoulis, A. *Probability, Random Variables, and Stochastic Processes,* 2nd ed. New York: McGraw-Hill, pp. 145–149, 1984.

Press, W. H., Flannery, B. P., Teukolsky, S. A., and Vetterling, W. T. "Moments of a Distribution: Mean, Variance, Skewness, and So Forth." §14.1 in *Numerical Recipes in FORTRAN: The Art of Scientific Computing,* 2nd ed. Cambridge, England: Cambridge University Press, pp. 604–609, 1992.

Kenney, J. F. and Keeping, E. S. "Moment-Generating and Characteristic Functions," "Some Examples of Moment-Generating Functions," and "Uniqueness Theorem for Characteristic Functions." §4.6–4.8 in *Mathematics of Statistics,* Pt. 2, 2nd ed. Princeton, NJ: Van Nostrand, pp. 72–77, 1951.

Weisstein, E. W. *Moment Generating Functions.* Wolfram Research, Inc., CRC Press LLC, 1999–2003.

Bendat, J. S., and Piersol, A. G. *Random Data: Analysis and Measurement Procedures*, 2nd ed. New York: Wiley-Interscience, John Wiley & Sons, 1986.

Bendat, J. S., and Piersol, A. G. *Engineering Applications of Correlation and Spectral Analysis*, 2nd ed. New York: Wiley-Interscience, John Wiley & Sons, 1993.

CHAPTER 8

Nonparametric Statistic and the Runs Test for Stationarity

8.1 INTRODUCTION

8.1.1 Conditions for a Random Process to Be Considered as Stationary

Recall that Chapter 7 contained a discussion on how the properties of a random process can be determined to be stationary by computing the ensemble averages at specific instants of time. A stationary random process was considered to be weakly stationary or stationary in the wide sense, when the following conditions were met:

(a) the mean, $\mu_x(t_1)$, termed the first moment of the random processes at all times t_i and

(b) the autocorrelation function, $R_x(t_1, t_1 + \tau)$, or joint moment between the values of the random process at two different times does not vary as time t varies.

That is, both the ensemble mean $\mu_x(t1) = \lim_{N\to\infty} \frac{1}{N} \sum_{k=1}^{N} x_k(t1)$ and the autocorrelation function $R_x(t_1, t_1 + \tau) = R_x(\lambda) = \lim_{N\to\infty} \frac{1}{N} \sum_{k=1}^{N} x_k(t_1)x(t_1 = \tau)$ are time invariant.

If all the higher moments and all the joint moments could be collected, one would have a complete family of probability distribution functions describing the processes. For the case where all possible moments and joint moments are time invariant, the random process is strongly stationary or stationary in the strict sense.

Unfortunately, it *is not* often possible to describe the properties of a stationary random process by computing time averages over specific sample functions in the ensemble, since many times an ensemble cannot be collected, but rather a single long recording of a signal is acquired. In this case, the long recording is divided into smaller segments and the data are tested for the Ergodic properties.

For example, the mean value $\mu_x(k)$ and the autocorrelation function $R_x(\lambda, k)$ of the kth sample function can be obtained with (8.1) and (8.2).

$$\mu_x(k) = \lim_{T \to \infty} \frac{1}{T} \int_0^T x_k(t)dt \tag{8.1}$$

$$R_x(\lambda, k) = \lim_{T \to \infty} \frac{1}{T} \int_0^T x_k(t)\, x_k(t + \tau)dt \tag{8.2}$$

The random process $\{x(t)\}$ is **Ergodic** and Stationary if the mean value $\mu_x(k)$ and the autocorrelation function $R_x(\tau, k)$ of the short-time segments **do not** differ when computed over different sample functions. It should be noted that only stationary random processes can be **Ergodic**. Another way of thinking is "all Ergodic Random Processes are Stationary." For many applications, verification of weak stationarity justifies an assumption of strong stationary (Bendat and Piersol).

8.1.2   Alternative Methods to Test for Stationarity

So far, the methods described require integral functions and parametric statistical testing or moments describing the properties of the random process. A disadvantage in the use of parametric statistics is that the random data must be tested for independence and normality. By normality of the distribution, we mean that the data must be "Gaussian distributed." By far the most common method used is to test whether the data are Ergodic. In general, three assumptions are made, when random data are tested for Ergodicity. The basic assumptions are as follows.

1. That any given sample record will properly reflect the nonstationary character of the random process.

2. That any given sample record is very long compared to the lowest frequency component in the data, excluding a nonstationary mean. What is meant by this assumption is that the sample record has to be long enough to permit nonstationary trends to be differentiated from the random fluctuations of the time history.

3. That any nonstationarity of interest will be revealed by time trends in the mean square value of the data.

With these assumptions, the stationarity of random data can be tested by investigating a single long record, $x(t)$, by the following procedure:

(1) Divide the sample record into N equal time intervals where the data in each interval may be considered independent.

(2) Compute a mean square value (or mean value and variance separately) for each interval and align these samples values in a time sequence as follows: $\overline{X}_1^2, \overline{X}_2^2, \overline{X}_3^2, \ldots \overline{X}_N^2$.

(3) Test the sequence of mean square values for the presence of underlying trends or variations other than those due to expected sampling variations. The final test of the sample values for nonstationary trends may be accomplished in many ways. A *nonparametric* approach, which *does not* require knowledge of the sampling distributions of data parameter, is desirable [3 and 4]. One such test is the Runs Test. There are other nonparametric tests that may be used. These tests will be presented in the section on review of nonparametric statistics.

8.1.2.1 Runs Test

Because of the requirements in testing with parametric statistics, that is, independence and normality of the distribution, nonparametric statistics are often used. The advantage of the nonparametric test is that *a priori* knowledge regarding the distribution is not necessary.

Runs test considers a sequence of N observed values of a random variable x in which it was decided to use the mean square value $(\overline{X_i^2})$, where each observation is classified into one of two mutually exclusive categories, which may be identified simply as $+$ or $-$. For example,

$$x_i \geq \overline{X} \text{ is } + \quad \text{and} \quad x_i < \overline{X} \text{ is} -$$

Runs test for independence of data may be extended to two simultaneous sequences of data (x_i, y_i). For example, if two simultaneous sequence of two sets of measures x_i and y_i, $i = 1, 2, 3 \ldots, N$, then

$$x_i \geq y_i(+) \quad \text{or} \quad x_i < y_i(-).$$

Back to the single variable data set, let us hypothesize that the sequence of sample mean square values $(\overline{x}_1^2, \ldots, \overline{x}_N^2)$ is each independent sample values of a random variable with mean square value of ψ_x^2, where $R_x(0) = \psi_x^2$ is the autocorrelation function with zero time displacement $\lambda = 0$, as shown in (8.3).

$$\psi_x^2 = \lim_{T \to \infty} \frac{1}{T} \int_0^T x^2(t)dt \quad \text{or} \quad R_x(0) = \psi_x^2 \tag{8.3}$$

If the hypothesis is true, the variations in the sequence of sample values will be random and display no trend. Hence, the number of runs in the sequence, relative to say the median value, will be as expected for a sequence of independent random observations of the random variables as in Table 8.2. If the number of runs is significantly different from the expected number given in the table, then the hypothesis of stationarity would be rejected; otherwise, the hypothesis is accepted. The acceptance region for the hypothesis is

$$r_{\frac{N}{2}}; 1 - \alpha/2 < r \leq r_{\frac{N}{2}}; \alpha/2$$

Note that this testing procedure is independent of frequency bandwidth of the data or the averaging time used. The Runs test is *not* limited to a sequence of mean square values, since it works equally well on mean values, RMS values, standard deviations, mean absolute values, or any other parameter estimate. Furthermore, it is not necessary for the data under investigation to be free of periodicities, but the fundamental period must be short compared to the averaging time used to compute sample values.

The number of runs that occur in a sequence of observations gives an indication as to whether or not results are independent random observations of the same random variables. Specifically, if a sequence of N observations is independent observations of the same random variable, that is, the probability of a $(+)$ or $(-)$ result does not change from one observation to the next, then the sampling distribution of the number of runs in the sequence is a random variable r with a mean value as shown in (8.4) and the variance is given by (8.5).

$$\mu_r = \frac{2N_1 N_2}{N} + 1 \tag{8.4}$$

where N_1 is the number of $(+)$ observations and N_2 is the number of $(-)$ observations.

$$\sigma_r^2 = \frac{2N_1 N_2 (2N_1 N_2 - N)}{N^2 (N_1 - 1)} \tag{8.5}$$

For the case where $N_1 = N_2 = N/2$ then the mean and variance quations become (8.6 and 8.7.)

$$\mu_r = \frac{N}{2} + 1 \tag{8.6}$$

and

$$\sigma_r^2 = \frac{N(N-2)}{4(N-1)} \tag{8.7}$$

There distribution function is given by Table 8.2.

$$r_\eta \text{ is, } \quad \text{where } \eta = \frac{N}{2} = N_1 = N_2, \quad r_\eta; 1 - \frac{\alpha}{2} \quad \text{and} \quad r_n; \frac{\alpha}{2}$$

In any case, a sequence of plus and minus observations is obtained from which the number of "RUNS" is determined. A "RUN" is defined as a sequence of identical observations that is followed and preceded by a different observation or no observation at all. There are tables for the distribution function of runs. Runs test can be used to evaluate data involving the question of testing a single sequence of observations for independence.

TABLE 8.1: Mean Squared Values of the Segments $N = 20$. Significance Level: $\alpha = 0.05$

$(\bar{x}_1^2)$	5.5	$(\bar{x}_6^2)$	5.7	$(\bar{x}_{11}^2)$	6.8	$(\bar{x}_{16}^2)$	5.4
$(\bar{x}_2^2)$	5.1	(7)	5.0	(12)	6.6	(17)	6.8
(3)	5.7	(8)	6.5	(13)	4.9	(18)	5.8
(4)	5.2	(9)	5.4	(14)	5.4	(19)	6.9
(5)	4.8	(10)	5.8	(15)	5.9	(20)	5.5

Let us look at an example where the random data are from a sample record of 480 seconds in duration. The sample record is divided into 20 segments of equal lengths or 24 seconds of durations with the mean squared values shown in Table 8.1.

8.1.2.2 Median $= 5.6$

Hence, mean squared values greater than the median value will be designated a $(+)$ and those values less than the median will be designated as $(-)$.

$$x_i \geq 5.6 = + \qquad x_i < 5.6 = -$$

Result

$$\frac{-- \; \pm \; -- \; \pm \; - \; \pm \; - \; \pm \pm \pm \; -- \; \pm \; - \; \pm \pm \pm \; -}{1 \quad 2 \quad 3 \quad 4\;5\;6\;7 \quad 8 \quad 9\;\;10\;11 \quad 12 \quad 13}$$

$r = 13$ runs are represented by the sequence of 20 observations.

$$\text{for} \quad \alpha = 0.05 \quad \text{or} \quad \frac{\alpha}{2} = 0.025$$

From Runs Distribution, (Table 8.2),

$$r_{10}; 1 - \frac{\alpha}{2} = r_{10} : 0.975 = 6$$

$$r_{10}; \frac{\alpha}{2} = r_{10} : 0.025 = 15$$

Then,

$$r_{10} : 0.975 = 6 < r = 13 < r_{10} : 0.025 = 15$$

TABLE 8.2: Percentage Points of Run Distribution Values of $r_{n;\alpha}$ Such That Prob $[r_n > r_n] = \alpha$, Where $n = N_1 = N_2 = N/2$.

$n = N/2$	α					
	0.99	0.975	0.95	0.05	0.025	0.01
5	2	2	3	8	9	9
6	2	3	3	10	10	11
7	3	3	4	11	12	12
8	4	4	5	12	13	13
9	4	5	6	13	14	15
10	5	6	6	15	15	16
11	6	7	7	16	16	17
12	7	7	8	17	18	18
13	7	8	9	18	19	20
14	8	9	10	19	20	21
15	9	10	11	20	21	22
16	10	11	11	22	22	23
18	11	12	13	24	25	26
20	13	14	15	26	27	28
25	17	18	19	32	33	34
30	21	22	24	37	39	40
35	25	27	28	43	44	46
40	30	31	33	48	50	51
45	34	36	37	54	55	57

(*cont.*)

TABLE 8.2: (*Continued*)

	α					
$n = N/2$	0.99	0.975	0.95	0.05	0.025	0.01
50	38	40	42	59	61	63
55	43	45	46	65	66	68
60	47	49	51	70	72	74
65	52	54	56	75	77	79
70	56	58	60	81	83	85
75	61	63	65	86	88	90
80	65	68	70	91	93	96
85	70	72	74	97	99	101
90	74	77	79	102	104	107
95	79	82	84	107	109	112
100	84	86	88	113	115	117

The hypothesis is accepted, since $r = 13$ falls within the range between 6 and 15; hence, the data are independent and ergodic; thus stationary. There is no reason to question independence of observations since there is no evidence of an underlying trend.

8.1.2.3 Summary of Ergodic Test Via the Runs Test

The statistical properties of a random process can be determined by computing ensemble averages at specific instances of time or from segments of a single long record. If the time averaged mean value $\mu_x(k)$ and the autocorrelation function $R_x(\lambda,k)$ are time invariant when computed over different sample functions, then the random process is Ergodic and Stationary. An easier approach in testing a single random signal for stationarity is to use the Nonparametric Runs Test.

8.2   REVIEW OF NONPARAMETRIC STATISTICS USED FOR TESTING STATIONARITY

8.2.1   Kolmogorov–Smirnov One-Sample Test

The Kolmogorov–Smirnov one-sample test is concerned with the degree of agreement between the distribution of a set of observed scores and some specified theoretical distribution [5 and 6]. The test method requires calculation of the maximum deviation, D, with (8.8).

$$D = \text{max}imum \left| F_0(X) - SN(X) \right| \tag{8.8}$$

where $F_0(X)$ equals the theoretical cumulative distribution under the null hypothesis, H_0. Then for any value of X, the value $F_0(X)$ is the proportion of cases expected to have scores equal to or less than X. $SN(X)$ is the observed cumulative frequency distribution of a random sample with N observations. D, maximum deviation, is the largest value of the difference between $F_0(X)$ and $SN(X)$.

If the maximum deviation, D, is greater than or equal to the table value, then the null hypothesis is rejected, and it is concluded that the data have trends and is, therefore, nonstationary.

An example of the Kolmogorov–Smirnov test when testing for stationarity of average rainfall data is shown in Table 8.3.

TABLE 8.3: Example of Kolmogorov-Smirnov Test

	x_1	x_2	x_3	x_4	x_5
f	2	3	3	1	1
$F_0(X)$	1/5	2/5	3/5	4/5	5/5
$SN(X)$	2/10	5/10	8/10	9/10	10/10
$F_0(X) - SN(X)$	0	1/10	2/10	1/10	0

Note: Calculated test value: $D = 0.2$; Table value: $D_{10,0.05} = 0.410$; and $\Delta x = 0.5$; such that, $x_1 = 0.5$, $x_2 = 1.0$, $x_3 = 1.5$, $x_4 = 2.0$, and $x_5 = 2.5$.

The test failed to reject H_0, since the calculated test value of D is less than the table value $(0.2 < 0.410)$. It is concluded that the data are stationary.

8.2.2   One-Sample Runs Test

For the One-Sample Runs test, the sample record must be illustrated by symbols; generally, by taking the median of the sample data and assigning all values above the median a symbol, n_1, and all values below the median another symbol, n_2. Once the values are represented by symbols, the number of runs can be determined. A run is defined as a succession of identical symbols.

For small samples, n_1 and $n_2 < 20$, the number of runs, r, is then compared to the value from the Runs table. If r falls within the critical values, then the null hypothesis is accepted, which infers failure to reject the null hypothesis, H_0.

If n_1 or n_2 is larger than 20, the table value is not used. For large samples, an approximation of the sampling distribution of r is the normal distribution. Therefore, a Z score is computed to test H_0, as shown in (8.9):

$$Z = \frac{r - \mu r}{\sigma r} = \frac{r - \left(\dfrac{2n_1 n_2}{n_1 + n_2} + 1\right)}{\sqrt{\dfrac{2n_1 n_2 (2n_1 n_2 - n_1 - n_2)}{(n_1 + n_2)^2 (n_1 + n_2 - 1)}}} \tag{8.9}$$

The calculated Z value from (8.9) is then compared to values in the Normal Distribution Table, which shows the one-tailed probability. For a two-tailed test, double the probability value (p) given in the table. If the p associated with the observed value of r is less than or equal to the significant level (α), then the null hypothesis H_0 is rejected and it is concluded that the data are nonstationarity.

8.2.3   Testing for Ergodicity

8.2.3.1   Wilcoxon-Matched Pairs Signed-Ranks Test

The Wilcoxon-matched pairs test is a powerful test because it considers relative magnitude and direction of differences. The process steps are as follows:

TABLE 8.4: Example of Wilcoxon-Matched Pairs Signed-Ranked Test

INTERVAL	DIFFERENCE	RANK OF DIFFERENCE	RANK WITH LESS FREQUENT SIGN
1	2	3	
2	−1	−1.5	1.5
3	3	4	
4	0		
5	−1	−1.5	1.5
6	5	5	
7	6	6	
	$N = 6$		$T = 3.0$

1. Take signed difference for each matched pair.

2. Rank differences without respect to sign; if there is a tie, assign the average of the tied ranks.

3. Affix "+" or "−" to rank according to the difference.

4. Determine T, where T is the smaller of the sums of like ranked signs.

5. Determine N, where N is the number of differences having signs (zeroes do not count).

6. If $N < 25$, then the critical value is obtained from the "Table of Critical Values of T in the Wilcoxon-Matched Pairs Signed-Ranks Test." If the observed T is less than or equal to the table value, then the null hypothesis, H_0, is not accepted (rejected).

7. If $N > 25$, then the Z value must be calculated from (8.10) and the table of Critical Values of T used.

$$Z = \frac{T - \frac{N(N+1)}{4}}{\sqrt{\frac{N(N+1)(2N+1)}{24}}} \qquad (8.10)$$

To test for ergoticity of a process, we must determine the ensemble average and time average for each interval. The ensemble average and time average are then paired for each interval and their differences computed. The results of an example are given in Table 8.4.

The value from the table at $T_{6,0.05}$ is 0. Since the calculated value of 3.0 is not less than or equal to zero (3.0 is not <0.0), the null hypothesis, H_0, is accepted (failure to reject H_0) and it is concluded that the ensemble average and time average do not differ; thus, the process is Ergodic.

For more detailed information on nonparametric statistical procedures, see Hollander and Wolfe [1] and Siegel [2].

8.3 REFERENCES

1. Hollander, M., and Wolfe, D. A. *Nonparametric Statistical Methods*. New York: John Wiley and Sons, 1973.

2. Siegel, S. *Nonparametric Statistics*. New York: McGraw-Hill Book Company, 1956.

3. Au, T., Shane, R. M., and Hoel, L. A. *Fundamentals of Systems Engineering: Probabilistic Models*. Reading, MA: Addison-Wesley Publishing Company, 1972.

4. Brown, R. G. *Introduction to Random Signal Analysis and Kalman Filtering*. New York: John Wiley and Sons, 1983.

5. O'Flynn, M. *Probabilities, Random Variables, and Random Processes*. New York: Harper & Row, Publishers, 1982.

6. Panter, P. F. *Modulation, Noise, and Spectral Analysis*. New York: McGraw-Hill Book Company, 1965.

CHAPTER 9

Correlation Functions

In the general theory of harmonic analysis, an expression of considerable importance and interest is the correlation function given in (9.1).

$$\frac{1}{T} \int_{-T_{1/2}}^{T_{1/2}} f_1(t) f_2(t + \tau) dt \tag{9.1}$$

For the case of periodic functions where $f_1(t)$ and $f_2(t)$ have the same fundamental angular frequency ω_1 and where τ is a continuous time of displacement in the range $(-\infty, \infty)$, independent of t (the ongoing time).

One important property of the correlation expression is the fact that its Fourier Transform is given by (9.2), where $f_1(t)$ has the complex spectrum $F_1(n)$, and $f_2(t)$ has $F_2(n)$.

$$\overline{F_1}(n) F_2(n) \tag{9.2}$$

The bar $(\overline{F_1}(n))$ indicates the complex conjugate of the quantity (function) over which it is placed. To establish this fact, the Fourier expansion of $f_1(t)$ and $f_2(t)$ can be expressed as (9.3) and (9.4).

$$f_1(t) = \sum_{n=-\infty}^{\infty} F_1(n) e^{jn\omega_1 t} \tag{9.3}$$

$$f_2(t) = \sum_{n=-\infty}^{\infty} F_2(n) e^{jn\omega_1 t} \tag{9.4}$$

where the complex spectrums are given by (9.5) and (9.6), respectively.

$$F_1(n) = \frac{1}{T_1} \int_{-T_{1/2}}^{T_{1/2}} f_1(t) e^{-jn\omega_1 t} dt \tag{9.5}$$

and

$$F_2(n) = \frac{1}{T_1} \int_{-T_{1/2}}^{T_{1/2}} f_2(t) e^{-jn\omega_1 t} dt \qquad (9.6)$$

However, it should be noted from (9.1) that the original equation is not for $f_2(t)$ but rather for $f_2(t + \tau)$. Thus, it was found that caution must be taken in the manner that the correlation function is expressed:

$$\frac{1}{T} \int_{-T_{1/2}}^{T_{1/2}} f_1(t) f_2(t + \tau) dt = \frac{1}{T} \int_{-T_{1/2}}^{T_{1/2}} f_1(t) dt \sum_{n=-\infty}^{\infty} F_2(n) e^{jn\omega_1(t+\tau)} \qquad (9.7)$$

The manner in which the right-hand side of the equation is written indicates that summation with respect to n comes first and is followed by integration with respect to t, as shown in (9.7).

At this point, if the order of summation and integration is inversed, the result would be given by (9.8):

$$\frac{1}{T_1} \int_{-T_{1/2}}^{T_{1/2}} f_1(t) f_2(t + \tau) dt = \sum_{n=-\infty}^{\infty} F_2(n) e^{jn\omega_1 \tau} \frac{1}{T_1} \int_{-T_{1/2}}^{T_{1/2}} f_1(t) e^{jn\omega_1 t} dt \qquad (9.8)$$

Note that in (9.8), the exponent in the second term (the integral) has a positive, $+j$, rather than a negative, $-j$, as shown in the integral ($\int dt$) of (9.5), where the last integral is recognized as the complex conjugate of $F_1(n)$. Hence, the result yields (9.10),

$$\frac{1}{T_1} \int_{-T_{1/2}}^{T_{1/2}} f_1(t) f_2(t + \tau) dt = \sum_{n=-\infty}^{\infty} [\overline{F_1(n)} F_2(n)] e^{jn\omega_1 \tau} \qquad (9.10)$$

which results in the general form of the inverse Fourier Transform (9.11) for a periodic function of fundamental angular frequency ω_1 and complex spectrum $\overline{F_1(n)} F_2(n)$ of (9.10).

$$f(t) = \sum_{n=-\infty}^{\infty} F(n) e^{jn\omega_1 t} \qquad (9.11)$$

However, note that (9.10) is in terms of delays, τ, instead of time, t, as in (9.11). Then by applying the definition of the Fourier Transform of the periodic function $f(t)$ as in (9.12),

$$F(n) = \frac{1}{T_1} \int_{-T_{1/2}}^{T_{1/2}} f(t)e^{-jn\omega_1 t}\,dt \quad \text{for} \quad n = 0, \pm 1, \pm 2, \pm \ldots \ldots \tag{9.12}$$

and substituting the convolution function for $f(t)$ as in (9.13):

$$\frac{1}{T} \int_{-T_{1/2}}^{T_{1/2}} f_1(t) f_2(t+\tau)\,dt = f(t) \tag{9.13}$$

the complex spectrum $\overline{F_1}(n)F_2(n)$ becomes (9.14):

$$\overline{F_1}(n)F_2(n) = \frac{1}{T_1} \int_{-T_{1/2}}^{T_{1/2}} e^{-jn\omega_1\tau}\,d\tau \int_{-T_{1/2}}^{T_{1/2}} f_1(t)f_2(t+\tau)\,dt \tag{9.14}$$

Again, the right integration is preformed first with respect to t and the resulting function of τ is then multiplied by $e^{-jn\omega_1\tau}$. Then, the product is integrated with respect to τ in order to obtain a function of n.

Thus the equations in (9.15) are Fourier Transforms of each other.

$$\frac{1}{T_1} \int_{-T_{1/2}}^{T_{1/2}} f_1(t)f_2(t+\tau)\,dt \quad \text{and} \quad \overline{F_1}(n)F_2(n) \tag{9.15}$$

This relationship is called the "Correlation Theorem" for periodic functions.

9.1   THE CORRELATION PROCESS

The correlation process involves three important operations:

(1) The periodic function $f_2(t)$ is given as delays or time displacement, τ.

(2) The displacement function is multiplied by the other periodic function of the same fundamental frequency.

(3) The product is averaged by integration over a complete period.

These steps are repeated for every value τ in the interval $(-\infty, \infty)$ so that a function is generated. In summary, the combination of the three operations—displacement, multiplication, and integration—are termed "Correlation," whereas folding, displacement, multiplication, and integration are referred to as "Convolution," a totally different process. Correlation is not limited to periodic functions, as it can be applied to aperiodic and random functions.

9.1.1   Autocorrelation Function

Basic statistical properties of importance in describing single stationary random processes include autocorrelation functions and autospectral density functions. In the correlation theorem for periodic functions, it was found that under the special conditions, the two functions may be identical, $f_1(t) = f_2(t)$. In this case, the correlation equation becomes (9.16) and is referred to as the "Autocorrelation Function."

$$\frac{1}{T_1} \int_{-T_{1/2}}^{T_{1/2}} f_1(t) f_1(t + \tau) dt = \sum_{n=-\infty}^{\infty} \overline{|F_1(n)|}^2 e^{jn\omega_1\tau} \tag{9.16}$$

For the case in which $\tau = 0$, it means that the squared value of the function $f_1(t)$ is equal to the sum of the squares of the absolute values over the entire range of harmonics in the spectrum of the function. Because the expression on the left divides the integrated results, the autocorrelation at zero delay is referred to as the *Mean Squared Value* of the function. The autocorrelation function is often subscripted as φ_{11} in the time domain or as Φ_{11} in the frequency domain. The reciprocal relations expressed in (9.17) and (9.18) form the basis for the "Autocorrelation Theorem" for periodic functions.

Expressed in words, the theorem states that the autocorrelation function and the power spectrum of a periodic function are "Fourier" transforms of each other. Equation (9.17) is the time domain expression, stating that the autocorrelation is equal to the "Inverse Fourier" transform of the power spectrum.

$$\varphi_{11}(\tau) = \sum_{n=-\infty}^{\infty} \Phi_{11}(n) e^{jn\omega_1\tau} \tag{9.17}$$

Equation (9.18) is the frequency domain expression, stating that the "Fourier" transform of the autocorrelation function is equal to the power spectrum of the function.

$$\Phi_{11}(n) = \int_{-\frac{T_1}{2}}^{\frac{T_1}{2}} \varphi_{11} e^{-jn\omega_1\tau} d\tau \tag{9.18}$$

9.2   PROPERTIES OF THE AUTOCORRELATION FUNCTION

One of the properties of an autocorrelation function is that although it retains all harmonics of the given function, it discards all their phase angles. In other words, all periodic functions having the same harmonic amplitudes but differing in their initial phase angles have the same autocorrelation function or that the power spectrum of a periodic function is *independent* of the phase angles of the harmonics.

Another important property is that the autocorrelation function is an *Even Function* of τ. An even periodic function satisfies the expression given by (9.20).

$$\varphi_{11}(-\tau) = \varphi_{11}(\tau) \tag{9.20}$$

The proof is as follows: writing (9.16) as a function of a negative delay, $-\tau$, gives (9.21).

$$\varphi_{11}(-\tau) = \frac{1}{T_1} \int_{-T_{1/2}}^{T_{1/2}} f_1(t) f_1(t - \tau) dt \tag{9.21}$$

Using the change of variable, since $x = (t - \tau)$ or $x = (t + \tau)$, then the limits become $T_{1/2} - \tau$; $-T_{1/2} - \tau$, and $f_1(t - \tau)$ becomes $f_1(x + \tau - \tau) = f_1(x)$. Likewise, $f_1(t)$ becomes $f_1(t + \tau)$. Also since $\varphi_{11}(\tau)$ is a periodic function of period T_1, integration over the interval, $-T_{1/2} - \tau$ to $T_{1/2} - \tau$, is the same as $T_{1/2}$ to $-T_{1/2}$. The result is 9.22:

$$\varphi_{11}(-\tau) = \frac{1}{T_1} \int_{-T_{1/2}}^{T_{1/2}} f_1(t) f_1(t + \tau) dt \quad \text{or that} \quad \varphi_{11}(-\tau) = \varphi_{11}(\tau) \tag{9.22}$$

9.3   STEPS IN THE AUTOCORRELATION PROCESS

The autocorrelation process requires the five steps listed as follows:

1. Change of variable from time (t) to delay (τ)

2. Negative translation along the horizontal x-axis (N is the total number of delays)

3. Positive incremental translation along the horizontal x-axis (one delay)

4. Multiplication of the two functions, $f_1(\tau)$ and $f_2(t - \tau)$

5. Integration of area

Steps 3 through 5 are repeated until $2N$ delays are completed

The process might be synopsized as determining the area under the curves (product) as the folded function is slid along the horizontal axis. Note the direction of translation; the conventional notation is as follows:

a. Movement to the right for positive (+) time

b. Movement to the left for negative (−) time

At this point, let us go through an example to visualize the autocorrelation process. The function f_1 is shown in Fig. 9.1 as two repeated graphs. To simplify the example, the change of variable will not be preformed or shown.

The second step is to translate or move one of the functions in the negative direction without reversing the function. Figure 9.2 shows what the graph of the two functions would look like. Note that $x = 0$ is in the middle of the axis.

The next step is that of moving the translated function in a positive direction by 2 delays (from 0 to +2) along the x-axis as shown in Fig. 9.3. Note that the overlapping area is the area of both functions between 0 and +2 on the x-axis.

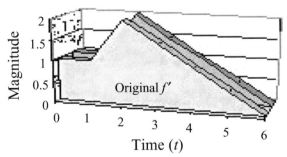

Autocorrelation first step

FIGURE 9.1: Starting the autocorrelation process. The function f_1 is shown as two separate 3-D objects

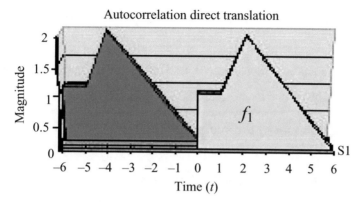

FIGURE 9.2: Step 2: Translation. Negative translation of function along the x-axis

9.4   NUMERICAL CORRELATION: DIRECT CALCULATION

9.4.1   Strips of Paper Method

One of the numerical methods is to use two strips of paper in the following steps:

1. Tabulate values of the function at equally spaced intervals on two strips of paper

2. Reverse tabulate one function; $f(-\lambda)$, which is equivalent to folding a function

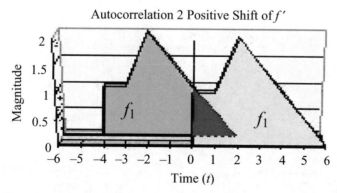

FIGURE 9.3: Positive translation. The second function is moved in a positive direction by 2 delays along the x-axis. Note the overlapping area of both functions between 0 and +2 on the x-axis

3. Place strips beside each other and multiply across the strips and write the products

4. Slide one strip, and repeat step 3, until there are no more products

5. Sum the rows

6. Multiply each sum by the interval width

9.4.2 Polynomial Multiplication Method

Another method is the polynomial multiplication method (Table 9.1). For this method, the values of the function are written twice in two rows; one function is placed above the other (reverse tabulated the second time the function is written) as in Table 9.1. It should be noted that the functions are aligned as in normal multiplication; from right to left. Reverse tabulation of the second function is important; otherwise, an

TABLE 9.1: Autocorrelation by Multiplication Method Correct Answer

			t							0	1	2	3	4	5	6
	f_1	D	T							1	1	2	1.5	1	0.5	0
	f_1	R	T							0	0.5	1	1.5	2	1	1
										1	1	2	1.5	1	0.5	0
									1	1	2	1.5	1	0.5	0	
								2	2	4	3	2	1	0		
							1.5	1.5	3	2.3	1.5	0.8	0			
						1	1	2	1.5	1	0.5	0				
					0.5	0.5	1	0.8	0.5	0.3	0					
				0	0	0	0	0	0	0						
	0	0.5	1.5	3.5	6.3	8	9.5	8	6.3	3.5	1.5	0.5	0			

Note: Original function (f_1); Time t ; DT is Direct Transcription; and RT is Reverse Transcription.

TABLE 9.2: Autocorrelation by Multiplication Method Incorrect Answer

			t			*0*	*1*	*2*	*3*	*4*	*5*	*6*
		*f*1	*D*	*T*		1	1	2	1.5	1	0.5	0
		*f*1	*D*	*T*		1	1	2	1.5	1	0.5	0
						0	0	0	0	0	0	0
					0.5	0.5	1	0.8	0.5	0.3	0	
				1	1	2	1.5	1	0.5	0		
			1.5	1.5	3	2.3	1.5	0.8	0			
		2	2	4	3	2	1	0				
	1	1	2	1.5	1	0.5	0					
1	1	2	1.5	1	0.5	0						
1	2	5	7	9	9	7.3	5	2.5	1	0.3	0	0

incorrect answer is obtained as shown in the calculations of Table 9.2. The number of terms in the results should equal twice the number of terms in the function. The final results of the correlation operation are obtained by summing the column and multiplying by the interval width. The graph of the correct calculation is shown in Fig. 9.4; note that the autocorrelation function is an even function (symmetrical about the y-axis).

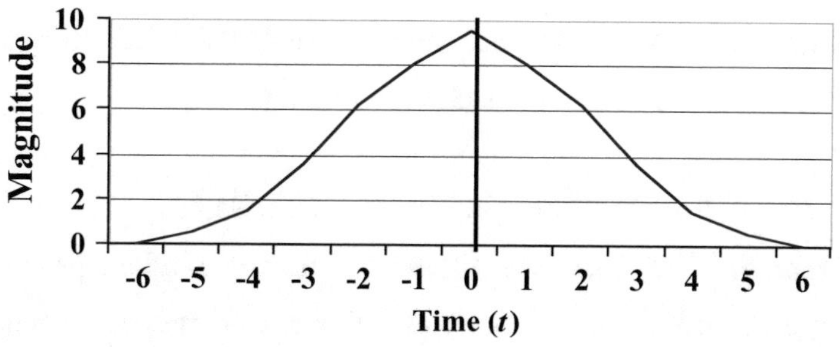

FIGURE 9.4: Correct calculation of autocorrelation. The graph shows the result of Table 9.1 calculations. Note that the autocorrelation function is an even function

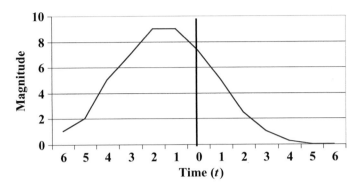

FIGURE 9.5: Incorrect autocorrelation results. Note the lack of symmetry in the graph of the incorrect calculation results of Table 9.2

For the Direct Numerical Method via polynomial multiplication, use the following steps:

1. directly tabulate values of the function in the second row with the increment of time above the function as in Table 9.1;

2. reverse tabulate the same function in the third row;

3. do normal polynomial multiplication (from right to left);

4. write the products in columns following normal multiplication procedures;

5. when there are no more products, sum the columns; and

6. multiply each sum by the interval width.

It is important to "Reverse tabulate" the second function; otherwise, the incorrect answer is obtained as shown in the calculations of Table 9.2. The graph of the incorrect results is shown in Fig. 9.5. Note the lack of symmetry.

9.5   CROSS-CORRELATION FUNCTION

For pairs of random data from two different stationary random processes, the joint statistics of importance include the cross-correlation and cross-spectral density functions. The cross-correlation theorem for periodic functions $f_1(t)$ and $f_2(t)$ applies to two different periodic functions, but with the stipulation that the functions must have the same fundamental frequency. If the condition of the same fundamental frequency is met, the cross-correlation function is defined as given by (9.23) for continuous data or by (9.24) for discrete data,

$$\varphi_{12}(\tau) = \frac{1}{T_1} \int_{-T_{1/2}}^{T_{1/2}} f_1(\tau) f_2(t + \tau) dt \tag{9.23}$$

$$r_{xy}(\tau) = \frac{1}{N} \sum_{i=1}^{n} x_i(t)\, y_t(t + \tau) = \overline{x(t)y(t + \tau)} \tag{9.24}$$

and cross-power spectrum of the functions as (9.25).

$$\Phi_{12}(n) = \overline{F_1(n)} F_2(n) \tag{9.25}$$

The subscripts 12 indicate that the cross-correlation involves $f_1(t)$ and $f_2(t)$ with the second numeral (2) referring to the fact that $f_2(t)$ has the displacement τ.

It is important to keep the numerals in the double subscripts in proper order; because, if we interchange the order of the subscripts, then by definition the resulting cross-correlation function is given by (9.26).

$$\varphi_{21}(\tau) = \frac{1}{T_1} \int_{-T_{1/2}}^{T_{1/2}} f_2(\tau) f_1(t + \tau) dt \tag{9.26}$$

Some text books will use the notation, R_{xy}, which is referred to as the correlation of x with y. The correlation of y with x is then expressed as (9.27). Note that the first subscript in either equation is the function without $(t + \tau)$. The preferred equation for the cross-correlation expression is (9.26).

$$R_{yx} = \int_{0}^{\infty} x(t + \tau) y(t) d\tau \tag{9.27}$$

The subscripts for the cross-spectrum are given in (9.28); note that the conjugated function is the first subscript.

$$\Phi_{21}(n) = \overline{F_2(n)} F_1(n) \tag{9.28}$$

The importance of keeping the numerals in the double subscript in the proper order

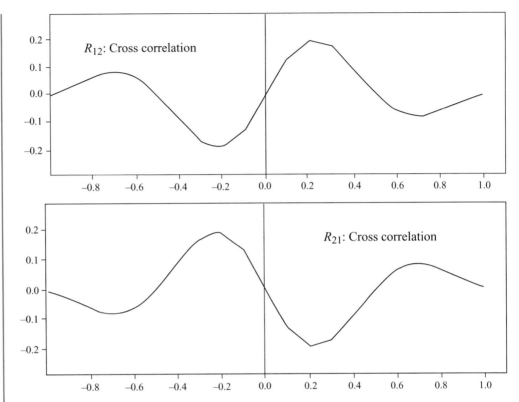

FIGURE 9.6: Cross-correlation images. The graphs show the comparison of the cross-correlation function $\Phi_{12}(n)[R_{12}]$ to its complex conjugate image $\Phi_{21}(n)[R_{21}]$

can not be overstressed. If the subscripts are interchanged, the results will not be what is expected. This interchange can be made by changing $+\tau$ in the definition to $-\tau$ and letting $x = t - \tau$, leading to the expression (9.29). The fact is that $\varphi_{12}(-\tau)$ is the mirror image of $\varphi_{21}(\tau)$ with respect to the vertical axis.

$$\varphi_{12}(-\tau) = \varphi_{21}(\tau) \tag{9.29}$$

The representation of the property in the frequency domain is given by (9.30).

$$\Phi_{12}(n) = \overline{\Phi}_{21}(n) \tag{9.30}$$

In summary, $\Phi_{12}(n)$ and $\Phi_{21}(n)$ are related to each other as complex conjugate quantities. Figure 9.6 shows the graphs comparing the cross-correlation function $\Phi_{12}(n)[R_{12}]$ to its complex conjugate image $\Phi_{21}(n)[R_{21}]$.

9.5.1   Properties of the Cross-Correlation Function

The cross-correlation function possesses several properties which are valuable in applications to signal and data processing:

1. $R_{xy}(\tau) = R_{xy}(-\tau)$ The cross-correlation is not an "Even Function."

2. $\left| R_{xy}(\tau) \right| \leq R_{xy}(0) R_{yx}(0) = \overline{x^2} * \overline{y^2}$ (the product of the RMS values of x and y)

3. If x and y are independent variables, then
    $R_{xy}(0) = R_{yx}(0) = \overline{x} * \overline{y}$ (the product of the mean values of x and y)

4. If x and y are both periodic with period $= T_0$, then R_{xy} and R_{yx} are also periodic with period $= T_0$.

5. The cross-power spectrum of x and y is computed as the Fourier Transform of the cross-correlation function.

9.5.2   Applications of the Cross-Correlation Function

1. An indication of the predictive power of x to y (or vice versa) is given by the magnitude of the correlation function. Delays of the effects of x on y are indicated by the value of t at which R_{xy} reaches a maximum.

2. The RMS and mean values of x or y can be estimated through the use of properties 2 and 3 given above, provided that the values of one function are known.

3. Periodicities in x and y can be detected in the cross-correlation function (Property 4).

4. The cross-power spectrum, given by the Fourier transform of the cross-correlation, is used in the calculation of the coherence of the functions.

C H A P T E R   1 0

Convolution

The convolution integral dates back to 1833. It is one of the methods used to determine the output response of a system to a specific input if the system transfer function is known. Convolution applies the superposition principal. Conceptually,

1. the input signal is represented as a continuum of impulses;

2. the response of the system to a single impulse is obtained;

3. the response of the system to each of the elementary impulses, representing the input, is computed; and then

4. the total input response is obtained by superposition.

Often signals are represented in terms of elementary linear basis functions, but not as a continuum of impulses. Therefore let us examine the representation of an arbitrary time function by a continuum of impulses. Let us begin with the approximation of a function $f(t)$ in the interval $-T$ to $+T$ by rectangles (rectangular pulses) whose height is equal to the function at the center of the rectangular pulse, and the width is ΔT (Fig. 10.1).

The equation describing the impulse representation of the continuous signal is given by (10.1).

$$f(t) = \int_{-T}^{+T} f(\tau)\delta(t - \tau)\,\partial\tau \quad \text{from} \quad -T < t < T \qquad (10.1)$$

Since the convolution is used in the time domain analysis of a filter response, let us

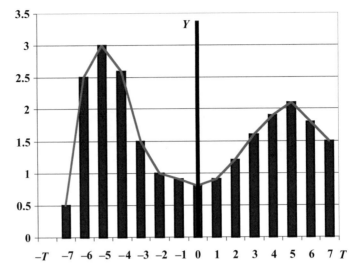

FIGURE 10.1: A signal represented in terms of a continuum of impulses in the interval, $-T$ to $+T$

examine how the convolution integral is used to obtain the system impulse response of a simple RC circuit or low pass filter as shown in Fig. 10.2.

If a single unit pulse of width ΔT and amplitude $1/\Delta T$ is applied, the output voltage increases exponentially toward the value $1/\Delta T$; when the pulse terminates at ΔT, the output voltage decays toward zero as shown in Fig. 10.3.

If the impulse is taken to its limit as $\Delta t \rightarrow 0$, then the impulse approaches a theoretical impulse; the *delta function*, δ, and the response to the impulse

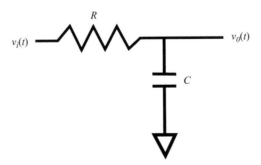

FIGURE 10.2: Low-pass RC filter circuit

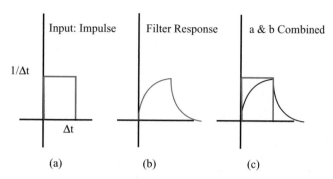

Input: Impulse	Filter Response	a & b Combined

(a)   (b)   (c)

FIGURE 10.3: Low-pass filter response. (a) It shows a single unity pulse of Δt duration and $1/\Delta t$ magnitude. (b) It shows the output response of the low-pass filter shown in Fig. 10.2 to the input pulse, (a). (c) It shows the combined input and output signals

becomes (10.2).

$$h(t) = \frac{1}{RC} e^{-\frac{t}{RC}} \qquad (10.2)$$

What Fig. 10.4 implies is that at the instant the impulse occurs a current flows in the circuit and is supposed to charge the capacitor to a voltage of 1/RC, instantaneously. However, the instantaneous charging that theoretically occurs in the instant that the impulse is applied is not physically possible because of the principle of conservation of momentum. As a note of information, the LaPlace transformation of the delta function, δ, is unity (1); therefore, the frequency domain is a better and easier method by which to analyze the response to an impulse function.

It is important to understand the time/frequency relationships that exist when functions are transformed into another domain. For example, the product (or *Modulation*)

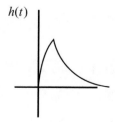

$h(t)$

FIGURE 10.4: Low-pass filter response to a delta function as $\Delta t \to 0$

of two functions in the time domain corresponds to convolution in the frequency domain as shown in (10.3).

$$f_1(t) \, f_2(t) = \frac{1}{2\pi} \int_{-\infty}^{\infty} F_1(\zeta) \, F_2(\omega - \zeta) \, \partial\zeta = \frac{1}{2\pi} F_1(\omega) \times F_2(\omega) \tag{10.3}$$

Whereas convolution in the time domain corresponds to product in the frequency domain as shown by (10.4.)

$$f_1(t) \times f_2(t) = \int_{-\infty}^{\infty} f_1(\lambda) \, f_2(t - \lambda) \, \partial\lambda \Leftrightarrow F_1(\omega) \, F_2(\omega) \tag{10.4}$$

Convolution is commutative for time-invariant systems are given by (10.5) and (10.6).

$$y(t) = \int_{-\infty}^{\infty} h(\lambda) x(t - \lambda) \, \partial\lambda \tag{10.5}$$

$$y(t) = \int_{-\infty}^{\infty} x(\lambda) h(t - \lambda) \, \partial\lambda \tag{10.6}$$

Often, the convolution operation is denoted by an asterisk as shown in (10.7).

$$f_1(t) \times f_2(t) = \int_{-\infty}^{\infty} f_1(\lambda) \, f_2(t - \lambda) \, \partial\lambda \tag{10.7}$$

The integration limits vary with the particular characteristics of the functions being convolved. Next, let us examine and discuss the manner of "Evaluation and Interpretation of Convolution Integrals and Numerical Convolution."

10.1   CONVOLUTION EVALUATION

Formally the convolution operation is given as in (10.8).

$$f_3 = f_1 \times f_2 = \int_{-\infty}^{\infty} f_1(\lambda) \, f_2(t - \lambda) \, \partial\lambda \tag{10.8}$$

FIGURE 10.5: Unit step function, $u(t)$. The step function is similar to throwing a switch

The value f_3 for any particular instant of time t is the area under the product of $f_1(\lambda)$ and $f_2(t - \lambda)$. To use the convolution technique efficiently, one should be able to visualize the two functions. If convolution is a new topic to the reader, then a quick review of wave form synthesis is essential for visualization. Let us begin with a *step function*, $u(t)$, which is initiated at $t = 0$ (Fig. 10.5).

Figure 10.6 shows the unit step function translated, $u(t - a)$, along the time axis, $+t$, by some length, $+a$, whereas Fig. 10.7 shows the unit step function translated backward along the time axis, $-t$, by some length, $-a$.

In addition, the unit step function may be rotated about the y-axis, which is referred to as *folding* about some axis. A folded (flipped) unit step function about the y-axis is written as, $u(-t)$, and is shown in Fig. 10.8.

Figure 10.9 shows the result of positive translation of a folded unit step function. Note that this function is written as, $u(a - t)$.

In addition to rotation about the y-axis, the unit step function may also be folded about the x-axis as shown in Fig. 10.10; note the negative sign in front of the unit step function, $-u(t)$.

Figure 10.11 shows the translation of a unit step function folded about the x-axis displaced by some time value, a. This step function is written as $-u(t - a)$; note the negative signs.

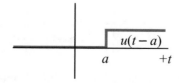

FIGURE 10.6: Positive translation. The unit step function of Fig. 10.5 is translated along the time axis, $+t$, by some length, $+a$

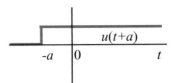

FIGURE 10.7: Negative translation. The unit step function, $u(t + a)$, translated backward along the time-axis, $-t$, by some length, $-a$

FIGURE 10.8: Folding. A unit step function rotated about the y-axis, $u(-t)$

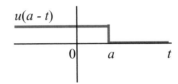

FIGURE 10.9: Positive translation of folded step function. Shown is the result of a positive translation in time (length, $+a$) of a folded unit step function, $u(a - t)$

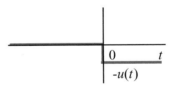

FIGURE 10.10: Folding about the x-axis. A unit step function is shown folded about the x-axis, $-u(t)$

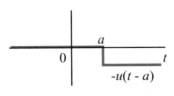

FIGURE 10.11: Folding and translation about the x-axis. A unit step function folded about the x-axis, t, then translated or displaced by some time value, $+a$

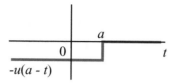

FIGURE 10.12: Combining the operations of folding and translating about both axes. The step function folded about both axes and displaced by $+a$ gives the unit step function expressed as $-u(a-t)$

Combining the operations of folding about the y-axis, then folding about the x-axis, and translating the step function by a positive displacement of $+a$ gives the results shown in Fig. 10.12. The unit step function is then expressed as $-u(a-t)$.

10.1.1 Real Translation Theorem

One can summarize the operation of translation along an axis by a well-known theorem which states that any function $f(t)$, delayed in its beginning to some time $t = a$, can be represented as a time-shifted function as in (10.9).

$$f(t-a)\,\mu(t-a) \tag{10.9}$$

where $\mu(t-a) = \begin{bmatrix} 1, -t \geq a \\ 0, -t < a \end{bmatrix}$

Proof of the time shift theorem is based in LaPlace Transform of the two functions, which is often referred to as the *Real Translation Theorem* or simply the *Shifting Theorem* as shown in the resulting (10.10).

$$L\left[\mu(t-a)\right] = \frac{e^{-as}}{s} \tag{10.10}$$

where, e^{-as} shifts the transform from $t = 0$ to $t = \infty$.

Let us return to the visualization of convolution. The function $f(\tau)$ should not be a problem since it is similar to the function $f(t)$, except for the change of independent variable from units of time t to units of delay τ. It is important to keep in mind that once the variable is changed from time t to delay τ that the unit of delay τ becomes the independent variable and that time t in the function $f(t-\tau)$ is the same as displacement a in the function $f(a-t)$.

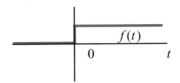

FIGURE 10.13: Simple unity step function

Before discussing the complete process of convolution, the reader should know that calculation of the convolution requires the following steps, which were previously described.

1. Change of variable from time, t, to delay, τ

2. Folding about the vertical axis

3. Translation

Let us look at an example, using the three steps to visualize and examine the process with a simple function, $f(t)$, as shown in Figs. 10.13–10.16.

The first step in the convolution process is the change of independent variable from a function of time, $f(t)$, to a function of delay, $f(\tau)$, as in Fig. 10.14.

The second step is to fold the step function about the y-axis, $f(-\lambda)$, as in Fig. 10.15.

The third step is to translate the step function along the x-axis some distance, t, as shown in Fig. 10.16. Note that the folded and translated step function is written as $f(t-\tau)$.

Let us examine another example; assume the function $f(t) = Ae^{-at}$. Figure 10.17 shows the function and the first step, which is the change of variable.

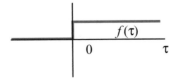

FIGURE 10.14: First step. Change the independent variable from a function of time, $f(t)$, to a function of delay, $f(\tau)$

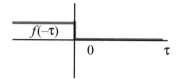

FIGURE 10.15: Second step. The step function is folded about the y-axis, $f(-\lambda)$

Figure 10.18 shows the folding and translation steps. Note that the translation is to the right. The function $f(t-\tau)$ can be thought of as $f\{-(\tau-t)\}$, then $f(-\tau')$ where τ' is $(\tau-t)$, which results in movement to the right (positive direction).

In the next example, illustrated in Fig. 10.19, the original does not begin at $t=0$, but is delayed by some magnitude a. The amount of displacement of the function is measured from the position of $f(-\tau)$ which is $f(t-\tau)$ at $t=0$. Note that the amount of delay is carried in the folding about the y-axis, which implies that the zero values must be accounted for in calculation of the convolution integral.

10.1.2   Steps in the Complete Convolution Process

The entire process of convolution requires the five steps listed as follows:

1. Change of variable

2. Folding about the vertical y-axis

3. Translation along the horizontal x-axis

4. Multiplication of the two functions, $f_1(\tau)$ and $f_2(t-\tau)$

5. Integration of areas

The process might be synopsized as determining the area under the curves (product) as the folded function is slid along the horizontal axis. Note the direction of translation; the conventional notation is as follows:

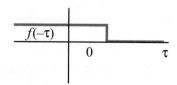

FIGURE 10.16: Third step. Translation of the function along the x-axis some distance, t

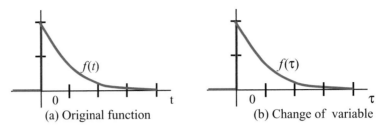

(a) Original function   (b) Change of variable

FIGURE 10.17: First step. (a) It shows the original function $f(t) = Ae^{-at}$. (b) It shows the change of variable for the function

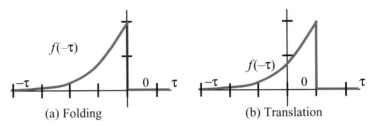

(a) Folding   (b) Translation

FIGURE 10.18: Next two steps. (a) It shows folding of the function about the y-axis and (b) it shows the translation step

a. Movement to the right for positive $(+)$ time

b. Movement to the left for negative $(-)$ time

Let us visualize an example, in which two functions of unity magnitude are to be convolved. Assume the two functions are given by (10.11) and (10.12).

$$f_1(t) = 1\mu(t) \tag{10.11}$$

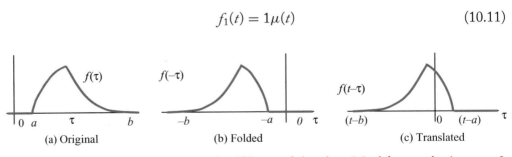

(a) Original   (b) Folded   (c) Translated

FIGURE 10.19: Another Example. It should be noted that the original does not begin at $t = 0$, but is delayed by some magnitude a. Figures 10.19(a) through (c) show that the amount of delay is carried through the Folding and Translation steps

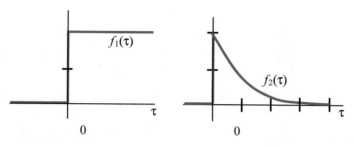

FIGURE 10.20: First step. Graphs of the two equations are shown after changing the independent variable

and

$$f_2(t) = e^{-at} \tag{10.12}$$

After changing the independent variable from time t to delay τ, the expressions are written as functions of τ, which are given in (10.13) and (10.14).

$$f_1(\tau) = \mu(\tau) \tag{10.13}$$

and

$$f_2(\tau) = e^{-a\tau} \tag{10.14}$$

Graphs of the two equations are shown in Fig. 10.20.

The next step is to fold one of the two functions about the y-axis. Either variable may be folded, since the results will be the same. In this case the step function was selected as the function to fold. Thus, $f_1(\tau) = \mu(\tau)$ becomes $f_1(-\tau) = \mu(-\tau)$ as shown in Fig. 10.21. Note that the second function, $f_2(\tau) = e^{-a\tau}$, remains the same.

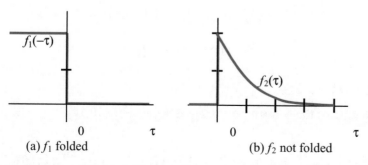

(a) f_1 folded   (b) f_2 not folded

FIGURE 10.21: Function f_1 Folded. The step function in Fig. (a) was folded while the second function was not folded

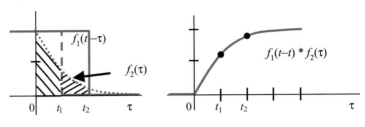

(a) Translation, product, and integration   (b) Resulting convolved function

FIGURE 10.22: Last three steps. The final steps of translation, product, and integration in the convolution process are shown in (a) with the resulting function shown in (b)

The final steps of translation, product, and integration in the convolution process are shown in Fig. 10.22(a) with the final convolution function being shown in Fig. 22(b). The reader should keep in mind that the integrated product is the same as the area under the product graph. In the first translation (t_1), the integrated quotient is shown as the area with the cross hatches slanting downward to the right. The new area of the integrated quotient for the second translation (t_2) is shown with the cross hatches slanting upward to the right. Note in the results, shown in Fig. 10.22(b), that the result at t_2 is the total area (both cross-hatched areas). The action of the functions f_1 and f_2 on each other is thought of as a smoothing or weighting operation.

Let us examine the convolution of a pulse and another function: assume the functions are defined as follows and are described by (10.15) and (10.16).

1. $f_1(t)$ is a rectangular pulse of height 3 and width 2
2. $f_2(t)$ is an exponential decay height 2 at $t = 0$

$$f_1(t) = 3\left[\mu(t) - \mu(t-2)\right] \tag{10.15}$$

$$f_2(t) = 2e^{-2t}\mu(t) \tag{10.16}$$

where $f_2(t) = 0$ for $t < 0$.

Evaluation of the convolution integral is given by (10.17).

$$f_3 = f_1(t) \times f_2(t) = 6\int_{-\infty}^{\infty}\left[\mu(t-\tau) - \mu(t-2-\tau)\right]e^{-2\tau}\mu(\tau)\,\partial\tau \tag{10.17}$$

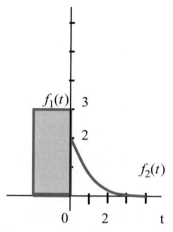

FIGURE 10.23: Second step: Folding. The figure shows the two functions after folding the pulse. It should be noted that there is no overlap of the functions when $t \leq 0$

Figure 10.23 shows the two functions after folding the pulse; note that at the start, when $t \leq 0$, there is no overlap and the resulting value of the convolution integral (f_3) at $t = 0$ is zero, ($f_3 = 0$).

Following the folding step, the exponential function, f_2, is multiplied by the pulse function, f_1, with a magnitude of 3, then the area between $0 < t < 1$ is integrated. The result of the integration is the first value of the convolution function, f_3, which is the shaded area in Fig. 10.24.

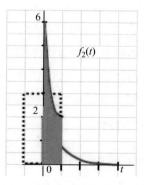

FIGURE 10.24: Result of integration. The first value of the convolution function, f_3, is shown by the shaded area in the figure

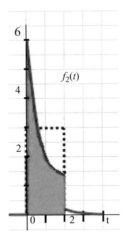

FIGURE 10.25: Result from the second translation. The second value after integration of the convolution function, f_3, with the limits between $0 < t < 2$, is shown by the shaded area in the figure

The next step is to repeat the translation step by moving the pulse function one increment to the right; hence, the limits of integration are changed to the interval of $0 < t < 2$, as given by (10.18).

$$f_3(t) = 6 \int_0^t e^{-2\tau} \partial \tau \qquad (10.18)$$

The result of the second translation and including the multiplication and integration steps is shown in Fig. 10.25. Note that the width of the area is equal to the width of the pulse function. As the pulse is translated to the next interval between $0 < t < 3$, it should be noted that the common area is reduced to the interval between $1 < t < 3$ (the width of the pulse). Hence, for this example, the area is smaller than the previous one. As the pulse is continuously translated, the value of the integrated area approaches zero.

The process repeats the three steps:

1. Translation along the horizontal x-axis

2. Multiplication of the two functions, $f_1(\tau)$ and $f_2(t - \tau)$

3. Integration of area bound by the limits of integration

In the example, for the next translations the limits are reset to the interval between $2 < t < \infty$ as shown in (10.19).

$$f_3(t) = 6 \int_{t-2}^{t} e^{-2\tau} \partial\tau \qquad (10.19)$$

10.1.3   Convolution as a Summation

In evaluation of the convolution integral by digital computer, it is necessary to replace the integral (10.20) with a finite summation.

$$v_2(t) = \int_{0}^{t} v_1(\tau) h(t - \tau) \partial\tau \qquad (10.20)$$

To accomplish this conversion, the infinitesimal $\partial\tau$ is replaced with a time interval of finite width T and τ is replaced with nT. The convolution integral is written as the summation for $KT \le t < (K+1)T$ as shown in (10.21).

$$v_2(t) = T \sum_{n=0}^{K} v_1(nT) h(t - nT) \qquad (10.21)$$

where K is one of the values of the integer n.

Equation (10.22) gives the convolution of (10.21) in expanded form.

$$
\begin{aligned}
v_2(t) = T[&v_1(0) h(t) + v_1(T)h(t - T) \\
&+ v_1(2T)h(t - 2T) + v_1 3T)h(t - 3T) + \cdots]
\end{aligned}
\qquad (10.22)
$$

Each term in (10.22) is the impulse response of magnitude of $v_1(nT)$ shifted along the t-axis and later is multiplied by T. Summation of the terms in the equation approximates the function. The accuracy of the approximation will depend on the magnitude of T. The smaller the value of T is, the better the approximation.

10.2   NUMERICAL CONVOLUTION

Numerical evaluation of the convolution integral is straightforward. When either function is of finite duration, there will be a finite number of terms in each summation.

However, when the functions are of infinite duration, the computations become lengthy as the overlapped area becomes very large. There are alternative methods that can be applied for simple problems or when a computer is available.

10.2.1   Direct Calculation

10.2.1.1   Strips of Paper Method

One of the numerical methods is to use two strips of paper and to follow the following steps:

1. tabulate values of the two functions at equally spaced intervals on two strips of paper;

2. reverse tabulate one function; $f(-\lambda)$, which is equivalent to folding a function;

3. place strips beside each other and multiply across the strips and write the products;

4. slide one strip, and repeat step three, until there are no more products;

5. sum the rows; and

6. multiply each sum by the interval width.

10.2.1.2   Polynomial Multiplication Method

Another method is the polynomial multiplication method (Table 10.1). For this method, the values of the two functions are placed in two rows one function above the other (without any reverse tabulation) as in Table 10.1. It should be noted that the functions are aligned to the left and not to the right as in normal multiplication. The alignment and multiplication from left to right serves as equivalence to folding of a function about the y-axis. The number of terms in the results should equal the sum of the number of terms in the two functions. The final results of the convolution operation are obtained by summing the columns and multiplying the resulting summations by the interval width.

At this point, it is worth discussing an area of possible confusion: that of establishing the proper relationship of the time variable to the sample values employed in the

TABLE 10.1: Convolution by Multiplication Method

	$t =$	1	2	3	4	5	6
	$f_1 =$	0.5	1.5	2.5	3	3	3
	$f_2 =$	9.5	8.5	7.5	6.5	5.5	
at $n = 19.5 \times f_1(n = 1..6)$		4.75	14.25	23.75	28.5	28.5	28.5
at $n = 18.5 \times f_1(n = 1..6)$			4.25	12.75	21.25	24.5	24.5	24.5
at $n = 17.5 \times f_1(n = 1..6)$				3.75	11.3	...	...	...
at $n = 16.5 \times f_1(n = 1..6)$					3.25	...	...	...
at $n = 15.5 \times f_1(n = 1..6)$						2.75	...	...
$f_1 \times f_2 = (\Sigma \times T)$		4.75	18.5	40.25	64.3	...	...	...

calculations. The question arises as to where to take the values from the time varying function within a fixed interval. Specifically, the question may be rephrased as "Where should the interval be located?" For example, a function, $f(t) = -1\lambda + 8$, can be sampled with some interval of time, T, in the following manner: Let $T = 2$ and the value be selected at midinterval, as in Fig. 10.26. The general procedure is to take the average value of the function $f(t)$ in the interval, kT, where k is the sample number.

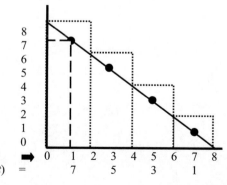

t ➡	0	1	2	3	4	5	6	7	8
$f(t) =$		7		5		3		1	

FIGURE 10.26: Graph A of the function, $f(t) = -1\lambda + 8$, with the interval $T = 2$. Note that the values of the sampled function are 7, 5, 3, and 1 at times $t = 1, 3, 5$ and 7, respectively

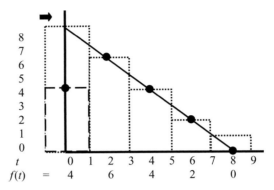

t		0	1	2	3	4	5	6	7	8	9
$f(t)$	=	4		6		4		2		0	

FIGURE 10.27: Graph B of the function, $f(t) = -1\lambda + 8$, with the interval $T = 2$. Note that the values of the sampled function are 4, 6, 4, 2, and 0 at times $t = 0, 2, 4, 6$, and 8, respectively

However, one may elect to use the interval between $-T/2 < 0 < +T/2$, which then yields slightly different answers as shown by Fig. 10.27. Note that in Fig. 10.27, the midpoints for the interval $T = 2$ are at $t = 1, 3, 5 \ldots$ etc.

Let us convolve the f_1 function with the values of the function $f(t) = -1\lambda + 8$, as shown in Table 10.1 and Fig. (10.26). The calculations and results of the convolution are given in Table 10.2 and are shown in Fig. (10.28). The results are labeled as A, since the second function is composed of the values from Fig. 10.26.

TABLE 10.2: Results A of Convolution with Fig. 10.26

	$t =$	1	2	3	4	5	6			
	$f_1 =$	0.5	1.5	2.5	3	3	3			
	$f_2 =$	7	5	3	1	0				
		3.5	10.5	17.5	21	21	21			
			2.5	7.5	12.5	15	15	15		
				1.5	4.5	7.5	9	9	9	
					0.5	1.5	2.5	3	2	1
						0	0	0	0	0	0
Col Sum		3.5	13	26.5	38.5	45	47.5	27	11	1	0
$f_1 \times f_2 = (\Sigma)(\Delta T)$		7	26	53	77	90	95	54	22	2	0

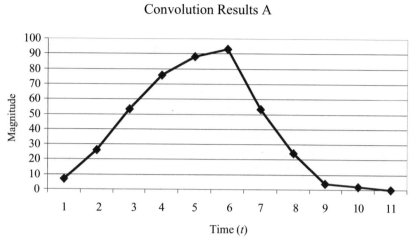

FIGURE 10.28: Graph of Results A

Repeating the convolution of the f_1 function with the values of the function, $f(t) = -1\lambda + 8$, as shown in Table 10.1 and Fig. 10.27. Note the values used in the calculations, which are given in Table 10.3. The results of the convolution are shown in Fig. 10.29. The results are labeled as B, since the second function is composed of the values from Fig. 10.27.

TABLE 10.3: Results B of Convolution with Fig. 10.27

	$t =$	1	2	3	4	5	6			
	$f_1 =$	0.5	1.5	2.5	3	3	3			
	$f_2 =$	4	6	4	2	0				
		2	6	10	12	12	12			
			3	9	15	18	18	18		
				2	6	10	12	12	12	
					1	3	5	6	6	
						0	0	0	0	0
Col Sum		2	9	21	34	43	47	36	18	0	0
$f_1 \times f_2 = (\Sigma)(\Delta T)$		4	18	42	68	86	94	72	36	0	0

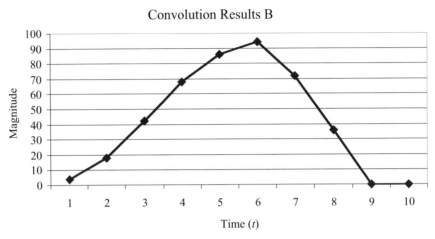

FIGURE 10.29: Graph of Results B

It should be noted that the procedures and results shown in Tables 10.2 and 10.3 and Figs. 10.28 and 10.29 are comparable, but the procedure that locates discontinuities along one of the steps in the approximation gives better results. In the example, the results labeled A (Table 10.2 and Fig. 10.28) are better for the discontinuity at $t = 0$.

10.3   CONVOLUTION ALGEBRA

Convolution is referred to as a "Linear Process" in the arithmetic sense; therefore, let us examine those mathematical properties that make convolution a linear process. Recall the mathematical properties of the "Real Numbering System" whose properties are

1. associative,

2. commutative, and

3. distributive.

Consider the convolution of two functions, one of the functions being the result of a previous convolution process. First, "Convolution is Associative," which means that the order of convolution is not important as indicated by (10.23).

$$f_1(t) \star [f_2(t) \star f_3(t)] = (f_1 \star f_2) \star f_3 = f_1 \star f_2 \star \overset{*}{f_3} \tag{10.23}$$

where $f_2 \star f_3 = \displaystyle\int_{-\infty}^{\infty} f_2(\lambda_1) f_3(t - \lambda_1)\, \partial\lambda_1$, and

$$f_1 \star [f_2 \star f_3] = \int_{-\infty}^{\infty} f_2(\tau)\, \partial\tau_2 \left[\int_{-\infty}^{\infty} f_2(\tau_1) f_3(t - \tau_1 - \tau_2)\partial\lambda_1 \right]$$

$$= \int \partial\tau_2 \int f_1(\tau_2) f_2(\tau_1) f_3(t - \tau_1 - \tau_2)\, \partial\tau_1$$

Second, "Convolution is Commutative," which means that the order of the functions being convolved is immaterial as shown in (10.24). The limits of integration will be different for different orders of the convolved functions.

$$f_1 \star f_2 = f_2 \star f_1 \tag{10.24}$$

The final property is that "Convolution is Distributive," which means that convolution of a function with the sum of two functions can be performed as shown in (10.25).

$$f_1 \star (f_2 + f_3) = f_1 \star f_2 + f_1 \star f_3 \tag{10.25}$$

An interesting property of convolution pertains convolving a function with theoretical delta functions, which results in the original function (see (10.26)). This property is the same as sampling.

$$z(t) \star \delta(t) = \int_{-\infty}^{\infty} z(\lambda)\delta(t - \lambda)\, \partial\lambda = z(t) \tag{10.26}$$

The convolution of a function with a unit step function is given by (10.27).

$$z(t) \times \mu(t) = \int_{-\infty}^{t} z(\lambda)\partial\lambda \tag{10.27}$$

If a function is convolved with the derivative of a delta function, the result is the derivative of the original function as shown by (10.28).

$$z(t) \star \delta'(t) = z'(t) \tag{10.28}$$

where the prime (') denotes d/dt. The first derivative of $\delta(t)$ is called a *doublet*.

The convolution of a step function with the derivative of a delta function will yield the delta function (10.29).

$$\mu(t) \star \delta'(t) = \delta(t) \tag{10.29}$$

It should be noted that the delta function is zero at all other values of time except at the instant of time, t, where the delta function exists. This relation is expressed mathematically as $\delta'(t - t_0) = 0$ for $t \neq t_0$ or as (10.30).

$$\int_{t_1}^{t_2} \delta(t - t_0)\, \partial t = 0 \tag{10.30}$$

For $t_1 < t_0 < t_2$

Other properties of convolution are of interest because of their use in checking validity of a particular computation. For example, in the convolution of two functions, as in (10.31), the area of a convolution product is equal to the product of the areas of the functions. This means that the area under $f_3(t)$ is equal to the area under the function, $f_1(t)$, times the area under the function, $f_2(t)$.

$$f_3(t) = \int_{-\infty}^{\infty} f_1(\tau) f_2(t - \tau)\, d\tau \tag{10.31}$$

Keep in mind that the area is computed by integrating over the interval, $-\infty < t < \infty$. Proof of this property follows in (10.32) through (10.36).

$$\int_{-\infty}^{\infty} f_3(t)\, dt = \int_{-\infty}^{\infty} \left[\int_{-\infty}^{\infty} f_1(\tau) f_2(t - \tau)\, d\tau \right] dt \tag{10.32}$$

$$[\text{Area\_under\_} f_3(t)] = \int_{-\infty}^{\infty} f_1(\tau) \left[\int_{-\infty}^{\infty} f_2(t - \tau)\, dt \right] d\tau \tag{10.33}$$

Rewriting (10.33), as moment generating functions, is shown in (10.34) and (10.35). Equation 10.36 shows that the moment about zero of the convolution function f_3 is equal to the product of the moments about zero of the two functions f_1 and f_2.

$$M_0(f_3) = \int_{-\infty}^{\infty} f_1(\tau)\left[\text{area\_under\_}f_2(t)\right] d\tau \qquad (10.34)$$

$$M_0(f_3) = A_{f_3(t)} = \left[\text{area\_under\_}f_1(t)\right] \bullet \lfloor A_{f_2(t)} \rfloor \qquad (10.35)$$

$$M_0(f_3) = M_0(f_1) \bullet M_0(f_2) \qquad (10.36)$$

Another useful relationship deals with the center of gravity (centroid) of the convolution in terms of the center of gravity of the factors (10.37). If you recall, the center of gravity of a waveform was defined in terms of the energy of the signal as t_0.

$$n = \frac{\int_{-\infty}^{\infty} tf(t)dt}{\int_{-\infty}^{\infty} f(t)dt} \qquad (10.37)$$

The nth moment of a function or waveform is defined by the general (10.38), and

$$M_n(f) = \int_{-\infty}^{\infty} t^n f(t)\, dt \qquad (10.38)$$

then the centroid is defined as given by (10.39).

$$n = \frac{M_1(f)}{M_0(f)} = \frac{M_1}{[\text{area\_under\_the\_waveform}]} \qquad (10.39)$$

The first moment of the convolution with infinite limits is given by (10.40).

$$M_1(f_3) = \int tf_3(t)\, dt = \int t\left[\int f_1(\tau) f_2(t - \tau)\, d\tau\right] dt \qquad (10.40)$$

The net result in moments is given by expression (10.41):

$$M_1(f_3) = M_1(f_1) M_0(f_2) + M_1(f_2) M_0(f_1) \qquad (10.41)$$

And since the moment about zero of the convolution function f_3 is equal to the product of the moments about zero of the two functions f_1 and f_2, $M_0(f_3) = M_0(f_1) \bullet M_0(f_2)$.

Then the ratio of the first moments to the moment about the origin can be expressed as (10.42) through (10.43).

$$\frac{M_1(f_3)}{M_0(f_3)} = \frac{M_1(f_1) M_0(f_2)}{M_0(f_1) M_0(f_2)} + \frac{M_1(f_2) M_0(f_1)}{M_0(f_1) M_0(f_2)} \tag{10.42}$$

After cancellation of common terms, the results are given in 10.43.

$$\frac{M_1(f_3)}{M_0(f_3)} = \frac{M_1(f_1)}{M_0(f_1)} + \frac{M_1(f_2)}{M_0(f_2)} \tag{10.43}$$

Rewriting (10.43) in short notation as $n_3 = n_1 + n_2$.

In the next step, the square of the centroid of the waveform is subtracted from the second moment of the waveform, which results in the square or the radius of gyration that corresponds to the variance in a probability distribution, as described by (10.44):

$$\sigma^2 = \frac{M_2(f)}{M_0(f)} - n^2 \tag{10.44}$$

where $M_2(f)$ is the second moment of a function $f(t)$.

The corresponding expression for the convolution operation is rewritten as (10.45)

$$\sigma_3^2 = \sigma_1^2 + \sigma_2^2 \tag{10.45}$$

The conclusion in probability is that the variance (of the sum of two independent random variables) equals the sum of their variances, since the probability density function of the sum is the convolution of the individual probability density functions.

10.3.1   Deconvolution

If "Convolution" can be thought of as corresponding to polynomial products, then the process of "Deconvolution" should correspond to polynomial division where one factor is known and the second must be recovered. Even though significant progress has been made in numerical methods, the process of deconvolution is very sensitive to errors and is not used in solution of most problems. There is a tendency for the residual (remainder) to become unstable and diverge in magnitude; hence, caution should be taken in using "Time-domain" deconvolution methods.

CHAPTER 11

Digital Filters

This chapter will review the concept of filters before presenting an overview of digital filters. It is not the intent of the chapter to enable the reader to write digital filter programs, but rather to present the engineer with the ability to use application software with filters and to choose the proper filter for specific biomedical applications. Let us start with the basic question, "What are Filters?" By definition, "Filters are circuits that pass the signals of certain frequencies and attenuate those of other frequencies."

11.1   CLASSIFICATION OF FILTERS

The most common classification or types of filters are as follows:

1. *Low-pass*: pass low frequencies; block high frequencies

2. *High-pass*: pass high frequencies; block low frequencies

3. *Band-pass*: pass a band of frequencies

4. *Band-reject*: attenuate a band of frequencies

5. *All-pass filter*: used for phase shifting

In addition filters are also categorized as follows:

1. Butterworth

   a. Low-pass filter

   b. High-pass filter, etc.

2. Chevbychev

   a. Low-pass filter

   b. High-pass filter, etc.

Other classifications that are also used in conjunction with the previous classification are

1. Voltage Controlled Voltage Source (VCVS),

2. Bi Quad,

3. Elliptical, and

4. Digital.

The most common digital filters fall into one of two classes:

1. Finite Impulse Response (FIR) and

2. Infinite Impulse Response (IIR)

Filters are generally described mathematically by their "Transfer Function" in the Time domain as differential equations, or in the Frequency domain as the ratio of two polynomials in "s-domain" or as $j\omega$ in radians. The order of a filter is defined by the degree of the denominator. The higher the order, the closer the filter approaches the ideal filter; however, the tradeoff an engineer must make is that higher order filters require more complex hardware, and consequently, are more expensive (costly).

11.1.1 Low-Pass Butterworth Filter

In many applications, the most common low-pass filter used is referred to as an "All-Pole Filter," which implies that the filter contained "All poles" and "No zeros" in the s-plane. In addition, the all pole, low-pass filter may be called a "Butterworth filter," which means that the filter has an "over damped" response. Figure 11.1 shows the ideal response of a low-pass Butterworth filter with its cutoff frequency at ω_c and the practical response for the same cutoff frequency. It should be noted that the low-pass Butterworth filter

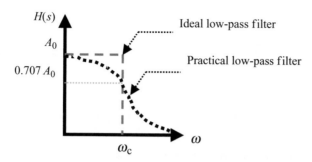

FIGURE 11.1: Butterworth low-pass characteristic transfer function response curves. The ideal low-pass filter is shown as the red trace, and the normal practical response trace is shown in black

magnitude monotonically decreases as frequency increases in both the band-pass and band-stop regions. The flattest magnitude of the filter is in the vicinity of $\omega = 0$, or DC, with its greatest error about the cutoff frequency.

The "Transfer Function" or describing equation for the Butterworth low-pass filter is given by (11.1) for the s-domain and rewritten in (11.2) in terms of frequency.

$$H(s) = \frac{K}{s+1} = \frac{A_{lp}}{s+j\omega} \tag{11.1}$$

$$H(j\omega) = \frac{A_{lp}}{\sqrt{1 + \left(\frac{\omega}{\omega_c}\right)^2}} \tag{11.2}$$

where A_{lp} = Low-pass gain, and $\frac{\omega}{\omega_c} = \omega_n$

The second-order transfer function for a low-pass Butterworth filter is given by (11.3):

$$H(s) = \frac{A_{lp}\omega_n^2}{s^2 + 2\zeta\omega_n s + b\omega_n^2}$$

where

$$b\omega_c^2 = \frac{1}{R_1 R_2 C_1 C_2} = \frac{1}{T_1 T_2} \tag{11.3}$$

11.1.2   Chebyshev low-pass Filter

The Chebyshev low-pass filter has an underdamped response, meaning that a filter of second order or higher order may oscillate and become unstable if the gain is too high

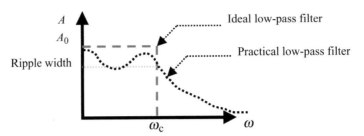

FIGURE 11.2: Chebyshev low-pass characteristic transfer function response curves. The ideal low-pass filter is shown as the dash-line trace, and the normal practical response trace is shown as the dotted-line trace. Note the ripple in the band-pass region

or the input is large. Hence, the term that the "filter rings," meaning it oscillates during it damping phase. The Chebyshev filter response best approaches the "Ideal filter" and is most accurate near the filter cutoff frequency, but at the expense of having ripples in the band-pass region. It should be noted that the filter is monotonic in the stop-band as is the Butterworth low-pass filter. Figure 11.2 shows the ideal response of a Chebyshev low-pass filter with its cutoff frequency at ω_c and the practical response for the same cutoff frequency.

The "Transfer Function" or describing equation for the Butterworth low-pass filter is given by (11.4) in terms of frequency.

$$|H(j\omega)| = \frac{A_{\text{lp}}}{\sqrt{1 + \varepsilon^2 C_n^2 \left(\frac{\omega}{\omega_c}\right)}} \tag{11.4}$$

where ε is a constant, and Cn is the Chebyshev polynomial, $Cn(x) = \cos[n \cos -1(x)]$
The ripple factor is calculated from (11.5).

$$RW = 1 - \frac{1}{\sqrt{1 + \varepsilon^2}} \tag{11.5}$$

Figure 11.3 shows the response of a third-order Chebyshev Voltage Controlled Voltage Source (VCVS) low-pass filter.

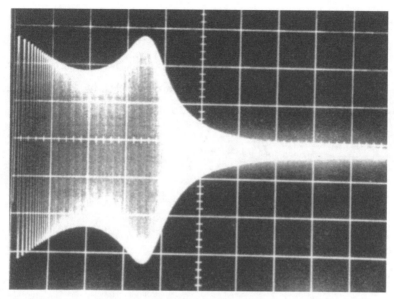

FIGURE 11.3: Chebyshev low-pass filter. Response of a third-order Chebyshev as shown by an oscilloscope

11.1.3   Butterworth High-Pass Filter

High-pass filters are designed to pass high frequencies and block low frequencies. As shown in Fig. 11.4, the Butterworth high-pass Filter magnitude monotonically decreases as frequency decreases, with the flattest magnitude in the high frequencies and the greatest error about the cutoff frequency.

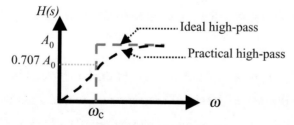

FIGURE 11.4: Butterworth high-pass transfer function characteristic curves. The ideal high-pass filter is shown as the square trace, and the normal practical response trace is shown in dashed-line black

The "Transfer Function" or describing equation for the first-order Butterworth high-pass filter is given by (11.6) for the s-domain and rewritten in (11.7) in terms of frequency. It should be noted that the high-pass filter is not an "All Pole Filter," since the equations show a zero or an s in the numerator of the equations. Hence, in the s-plane a zero exists at the origin, $s = 0$.

$$H(s) = \frac{Ks}{s + \frac{\omega_c}{b}} = \frac{A_{hp}s}{s + \frac{\omega_c}{b}} \tag{11.6}$$

$$H(j\omega) = \frac{A_{hp}s}{s + \frac{1}{R_1 C}} \tag{11.7}$$

where A_{hp} = high-pass gain.

11.1.4   2nd-Order Butterworth High-Pass Filter

The transfer function for a second-order Butterworth high-pass filter is given by (11.8):

$$H(s) = \frac{A_{hp}bs^2}{s^2 + \frac{a}{b}\omega_c s + \frac{\omega_c^2}{b}} \tag{11.8}$$

11.1.5   Band-Pass Filters

Band-pass filters may be of the Butterworth or of the Chebyshev class. Band-pass filters are designed to pass a band of frequencies of bandwidth, B, with the center of the band around a center frequency (ω_0 rad/sec) or $f_0 = \omega_0/2\pi$ Hz. Band-pass filter must designate two cutoff frequencies: a lower cutoff, ωL, and an upper cutoff, ωU.

The frequencies that are passed by the filter are those for which the transfer function, $H(s)$, gain is greater than or equal to $0.707A_0$ as shown in Fig. 11.5.

11.1.5.1   Quality Factor

Quality Factor (Q) is a measure of narrowness of the band-pass filter. By definition, the quality factor, Q, is the ratio of the center frequency, ω_0, to the bandwidth, $B = \omega U - \omega L$ in radians per second; hence, the ratio may be expressed as $Q = \omega_0/B$ or $Q = f_0/B$. If the quality factor, Q, is greater than 10, the band-pass region is considered to be narrow. The gain of the band-pass filter is the magnitude at the center frequency. The transfer function for the band-pass filter is given by (11.9). Band-pass filters must be of second

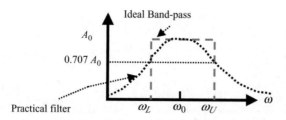

FIGURE 11.5: Band-pass filter transfer function characteristic curves. The ideal band-pass filter is shown as the dashed-line trace, and the normal practical response trace is shown as dotted-line trace

or higher order, since the filters require two slopes (high and low). Note that band-pass filters are not All-Pole filters, since the equation shows a zero in the numerator.

$$H(s) = \frac{A_{\mathrm{BP}}\omega_0 s}{s^2 + \frac{\omega_0}{Q}s + \omega_0^2} \tag{11.9}$$

11.1.6   Band-Reject or "Notch" Filters

Band-reject filters may be of the Butterworth or of the Chebyshev class. Band-reject filters are designed to attenuate or reject a band of frequencies of bandwidth, B, with the center of the band around a center frequency (ω_0 rad/sec) or $f_0 = \omega_0/2\pi$ Hz. As with the band-pass filter, it is necessary to designate the two cutoff frequencies for a band-reject filter; the lower cutoff, ωL, and the upper cutoff, ωU. The frequencies that are passed by the filter are those for which the transfer function, $H(s)$, gain is greater than or equal to $0.707A_0$ as shown in Fig. 11.6.

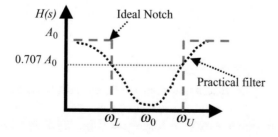

FIGURE 11.6: Band-reject filter transfer function characteristic curves. The ideal band-reject or notch filter is shown as the dashed-line trace, and the normal practical response trace is shown in dotted-line trace

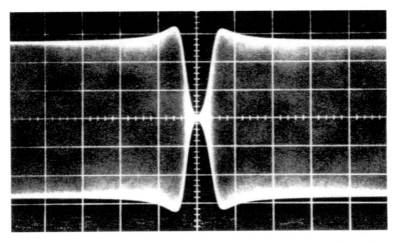

FIGURE 11.7: Chebyshev band-reject filter. Response of a fourth-order Chebyshev notch filter as shown by on an oscilloscope

The transfer function for the band-reject filter is given by (11.10). Band-reject filters must be of second or higher order, since the filters require two slopes (high and low). Note that band-reject filters are not All-Pole filters, since the equation shows a pair of zeros in the numerator (s^2).

$$H(s) = \frac{A_{BR}(s^2 + \omega_0^2)}{s^2 + \frac{\omega_0}{Q}s + \omega_0^2} \tag{11.10}$$

Figure 11.7 shows the response of a fourth-order Chebyshev band-reject (notch) filter. It should be noted that the maximum attenuation of the band-reject filter is the magnitude at the center frequency, ω_0.

11.2   DIGITAL FILTERS

11.2.1   Classification of Digital Filters

The classification of digital filters is similar to the classification of analog filters

1. Chebyshev filters: low-pass or high-pass (Figs. 11.8 and 11.9)

    a. Underdamped, can oscillate

    b. Ripple factor

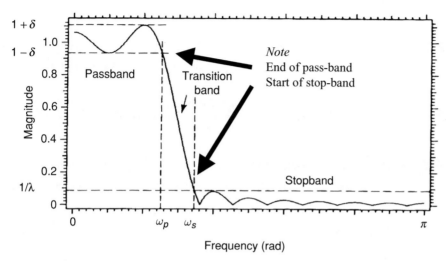

FIGURE 11.8: Low-pass filter. In specifying a digital filter, the end of pass-band and the stop-band parameters must be designated

2. Butterworth filters

    a. Overdamped

    b. Monotonic increasing or decreasing function

3. Elliptical filters

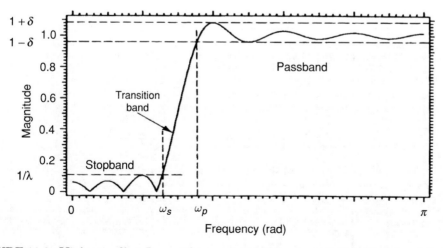

FIGURE 11.9: High-pass filter. In specifying a digital filter, the end of pass-band and the stop-band parameters must be designated. Note the stop-band, the transition band, and the pass-band

11.2.2   Infinite Impulse Response (IIR) Digital Filters

The Pole-zero transfer function of an Lth-order direct form Infinite Impulse Response Filter (IIR-filter) has the z-transform transfer function described by (11.11).

$$H(z) = \frac{B(z)}{A(z)} = \frac{b_o + b_{1Z}^{-1} + \cdots + b_{LZ}^{-L}}{1 + a_{1Z}^{-1} + \cdots + a_{LZ}^{-L}} \qquad (11.11)$$

where the zs are the "z-transformation."

General characteristics of the infinite impulse response filters are as follows:

a.   Very sharp cutoff

b.   Narrow transition band

c.   Low-order structure

d.   Low-processing time

There are several designs that can be used to implement (11.11), such as the direct, parallel, or cascade design. The "Direct" design is shown in Fig. 11.10, whereas a "Symmetric Lattice" of the IIR design is shown in Fig. 11.11.

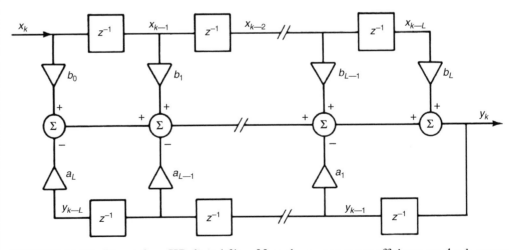

FIGURE 11.10: Direct-form IIR digital filter. Note the numerator coefficients are the b_L terms and the denominator coefficients are the a_L terms

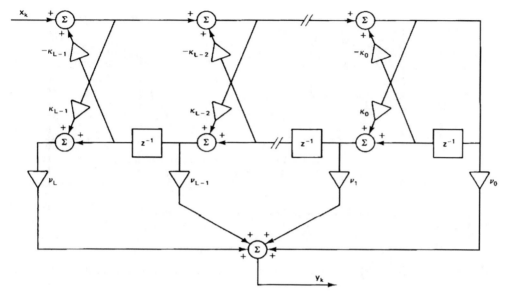

FIGURE 11.11: Symmetric IIR lattice digital filter

IIR filter also have a very narrow transition band, which would be desirable in band-pass and band-reject filters as shown in Figs. 11.12 and 11.13, respectively.

11.2.3   Finite Impulse Response (FIR) Digital Filters

The Finite Impulse Response (FIR) filter maybe derived from the IIR transfer function equation (11.11). If the denominator terms are unity, $A(z) = 1$, the equation for $H(z)$

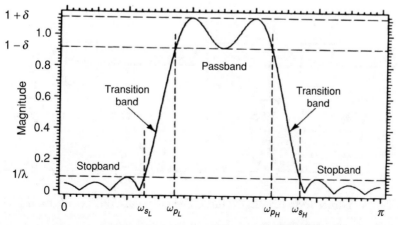

FIGURE 11.12: Band-pass filter. Note the narrow transition bands in the band-pass filter

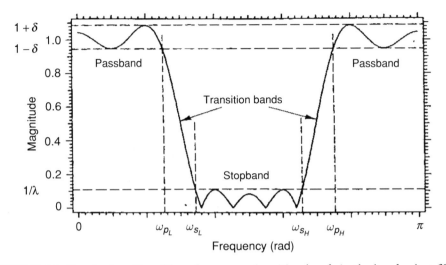

FIGURE 11.13: Band-reject filter. Note the narrow transition bands in the band-reject filter

in the IIR direct version becomes the FIR direct-form implementation equation, as in
(11.12).

$$H(z) = B(z) = b_0 + b_1 z^{-1} + \cdots + b_L z^{-L} \tag{11.12}$$

The difference equation corresponding to the FIR direct-form implementation equation
(11.12) is given by (11.13), and the "Direct" FIR design is shown in Fig. 11.14.

$$y_k = \sum_{n=0}^{L} b_n x_k - n \tag{11.13}$$

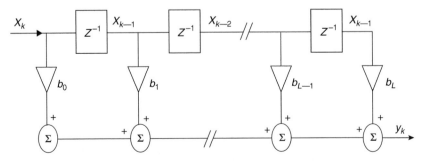

FIGURE 11.14: Direct-form FIR digital filter. The numerator coefficients are the b_L terms, and
since the a_L terms are unity there are no denominator coefficients

In summary, FIR filters should be used when phase is important or when flatness without ripples in the pass-band region is important in the analysis of signals. The general characteristics of the FIR filters are listed as follows:

a. many coefficients;

b. designed to have exactly linear phase characteristics; and

c. implementation via fast convolution.

C H A P T E R   1 2

Fourier Series: Trigonometric

The objective of this chapter is for the reader to gain a better understanding of the Basic Fourier Trigonometric Series Transformation and waveform synthesis. If we are interested in studying time-domain responses in networks subjected to periodic inputs in terms of their frequency content, the Fourier Series can be used to represent arbitrary periodic functions as an infinite series of sinusoids of harmonically (not harmoniously) related frequencies as shown in (12.1).

A signal $f(t)$ is periodic with period, T, if $f(t) = f(t + T)$ for all t, then the Fourier Trigonometric Series representation of the periodic function may be expressed as (12.1). It should be noted that both "cosine" and "sine" trigonometric terms are necessary and that an infinite number of terms may be necessary.

$$f(t) = a_0 + a_1 \cos \omega_0 t + a_2 \cos 2\omega_0 t + \cdots a_n \cos n\omega_0 t + b_1 \sin \omega_0 t$$
$$+ b_2 \sin 2\omega_0 t + \cdots b_n \sin n\omega_0 t + \cdots n = 1, 2, 3, \ldots \infty \qquad (12.1)$$

where $\omega_0 = \frac{2n\pi}{T}$; $T = \frac{1}{f_0}$; and n is the nth harmonic of the fundamental frequency, ω_0.

12.1 FOURIER ANALYSIS

Recall the chapter on basis functions. In taking the integral transform of a signal, the integral of the product of the signal with any basis function transforms the work from

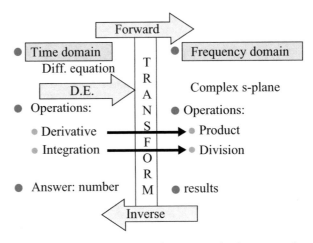

FIGURE 12.1: Transformation from the time domain to the frequency domain

one domain, that is, time domain, into a separate domain, that is, frequency domain with either the La Place or Fourier Basis functions. The reason for working in the transformed domain is to reduce the time-domain operations of integration or differentiation to less complex operations of division or multiplication, respectively.

There are numerous forms of the Fourier Transform. The most used forms include the following:

1. The Fourier Trigonometric Series, which was used prior to the advent of computer to analyze analog data, but had the major drawback of being very time consuming.

2. The Discrete Fourier Transform was used with digitized data, and is considered to be a subset of the LaPlace Transform; however, the Discrete Fourier Transform was also very time consuming (4 minutes to analyze 1024 samples of data).

3. The Fast Fourier Transform became the most popular method used to perform transformation into the frequency domain because it was very efficient and fast.

There are many other Fourier Transforms that will not be studied in this course.

The Fourier analysis consists of two basic operations:

1. The first operation is to determine the values of the coefficients $a_0, a_1, \ldots a_n$, and $b_1, \ldots b_n$.

2. The second operation is to decide how many terms to include in a truncated series such that the partial sum will represent the function within allowable error. It is not possible to calculate an infinite number of coefficients.

12.2   EVALUATION OF THE FOURIER COEFFICIENTS

In evaluating the coefficients, the limits of integration must be set. Since the Fourier Basis functions are orthogonal over the interval t_0 to $(t_0 + T)$ for any t_0, we often use the value $t_0 = 0$ or $t_0 = {}^{-T}/2$ with the understanding that any period may be used as the period of integration. Thus, we replace the interval of the integral from, t_0 to $(t_0 + T)$, with the limits of integration from, 0 to T.

The next step would be to calculate the a_0 term, which is simply the average value of $f(t)$ over a period; a_0 is often referred to as the DC value of a sinusoid over N complete cycles in the period. Next the complex coefficients, a_n and b_n, are evaluated with the expressions in (12.2), (12.3), and (12.4). To obtain the a_n terms, which constitute the "real-part" of the complex Fourier coefficient, the signal or function is multiplied by the corresponding cosine term, integrated, and then normalized by the fundamental period, T, as shown in (12.3). In a similar manner, to obtain the b_n terms, which constitute the "imaginary-part" of the complex Fourier coefficient, the signal or function is multiplied by the corresponding sine term, integrated, and then normalized by the fundamental period, T, as shown in (12.4).

DC Value of $f(t)$: $$a_0 = c_0 = \frac{1}{T}\int_{-T/2}^{T/2} f(t)dt = \frac{1}{T}\int_{t_0}^{t_0+T} f(t)dt = \frac{1}{T}\int_{0}^{T} f(t)dt$$

$$(12.2)$$

Real Term: $$a_n = \frac{2}{T}\int_{0}^{T} f(t)\cos n\omega_0 t\, dt \text{ or } \int_{-T/2}^{T/2} \qquad (12.3)$$

Imaginary term: $$b_n = \frac{2}{T}\int_{0}^{T} f(t)\sin n\omega_0 t\, dt \qquad (12.4)$$

12.3   EQUIVALENT FORM OF THE FOURIER TRIGONOMETRIC SERIES

The shorthand notation for writing the Fourier series is given by (12.5).

$$f(t) = a_0 + a_1 \cos \omega_0 t + a_2 \cos 2\omega_0 t + \cdots a_n \cos n\omega_0 t + \cdots b_1 \sin \omega_0 t$$
$$+ b_2 \sin 2\omega_0 t + \cdots b_n \sin n\omega_0 t + \cdots$$
$$f(t) = a_0 + \sum_{n=1}^{\infty} (a_n \cos n\omega_0 t + b_n \sin n\omega_0 t) \tag{12.5}$$

However, an equivalent expression is to combine the complex coefficients, a_n and b_n, and rewrite the coefficients in polar-coordinate terms of magnitude and phase as in (12.6).

$$c_n \cos(n\omega_0 t + \theta_n) = a_n \cos n\omega_0 t + b_n \sin n\omega_0 t \tag{12.6}$$

where $c_n = \sqrt{a_n^2 + b_n^2}$ and $\theta = \tan^{-1} \frac{b_n}{a_n}$

Then, the Fourier Trigonometric Series representation of the function may be written in the expanded form as in (12.7) or in the compressed form as in (12.8).

$$f(t) = c_0 + c_1 \cos(\omega_0 t + \theta_1) + \cdots c_n \cos(n\omega_0 t + \theta_n) + \cdots \tag{12.7}$$

where c_n is the **magnitude**, and θ_n is the **phase** of the nth harmonic.

$$f(t) = c_0 + \sum_{n=1}^{\infty} c_n \cos(n\omega_0 t + \theta_n) \tag{12.8}$$

12.4   SYMMETRICAL PROPERTIES AS RELATED TO FOURIER COEFFICIENTS

12.4.1 Even Waveform Symmetries

In working with the Fourier Series, it is often advantageous to take advantages of the symmetrical properties of trigonometric functions or other wave forms. If the function $f(t)$ satisfies the condition, $f(t) = f(-t)$, then the function is said to be *even*, as shown in Figs. 12.2 and 12.3. The function will contain only the Fourier coefficients; DC, a_0, and $a_n \cos n\omega_0 t$ terms. Note that the cosine function in Fig. 12.2 is an "even" function as is the square wave function in Fig. 12.3.

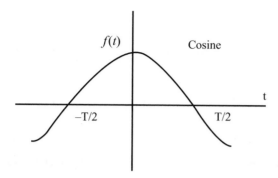

FIGURE 12.2: An even function. The cosine is an even function, since the function is symmetrical about the y-axis, that is, $f(t) = f(-t)$

12.4.2 ODD Waveform Symmetries

Similarly, if $f(t)$ satisfies the condition $f(t) = -f(-t)$, then the function is said to be *odd*, as shown in Fig. 12.4. The function will contain only the Fourier coefficients $b_n \sin n\omega_0 t$ terms. Note that the functions are folded symmetrically about both y- and x-axes. The negative sign before the time, t, folds the function about the y-axis whereas the negative sign before the function, f, inverts the function about the x-axis.

Another useful property to know is the "half-wave" symmetry. If the function $f(t)$ satisfies the condition $f(t) = -f(t \pm \frac{T}{2})$, then the function is said to have *half-wave* symmetry, as shown in Fig. 12.5. The function will contain only the odd complex Fourier coefficients, a_n and b_n, for $n = 1, 3, 5, 7, \ldots$, etc.

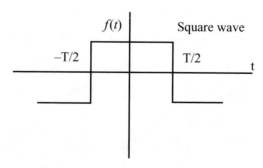

FIGURE 12.3: Another even function. The square wave function is an even function, since the function is symmetrical about the y-axis, i.e., $f(t) = f(-t)$

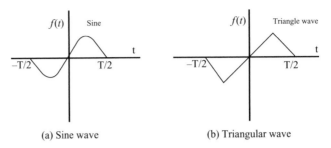

(a) Sine wave   (b) Triangular wave

FIGURE 12.4: Odd functions. The sine wave (a) and the triangular wave are considered to be odd functions since they satisfy the condition $f(t) = -f(-t)$

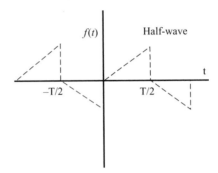

FIGURE 12.5: Half-wave symmetric function. This triangular function is considered to have half-wave symmetry since it satisfies the condition $f(t) = -f(t \pm \frac{T}{2})$

The relationship between the properties of symmetry and the Fourier Series are summarized in Table 12.10 and as follows:

1. Even functions contain only a_0 and $a_n \cos n\omega_0 t$ terms.

2. Odd functions contain only $b_n \sin n\omega_0 t$ terms.

3. Half-wave symmetry functions contain only the odd complex Fourier coefficients, a_n and b_n terms, where $n = 1, 3, 5, 7, \ldots$, etc.

Example Problem: What is the Fourier Series of the function, $v(t)$, in Fig. 12.6?

The function $v(t)$ is given below.

$$v(t) = \begin{cases} V, & 0 < t < T/4 \\ -V, & T/4 < t < 3/4\,T \\ V, & 3/4\,T < t < T \end{cases} \qquad (12.9)$$

TABLE 12.10 SUMMARY OF SYMMETRY TABLES

TABLE A: Waveform Symmetry

SYMMETRY	CONDITION	ILLUSTRATION	PROPERTY
Even	$f(t) = f(-t)$		cosine terms only
Odd	$f(t) = -f(-t)$		sine terms only
Half-wave	$f(t) = -f(t \pm T/2)$		odd n only

TABLE B: Fourier Transform of Symmetrical Waveforms

PROPERTY	a_0	$a_n (n \neq 0)$	b_n
cosine terms only	*	$\dfrac{4}{T}\displaystyle\int_0^{T/2} f(t) \cos n\omega_0 t\, dt$	0
sine terms only	0	0	$\dfrac{4}{T}\displaystyle\int_0^{T/2} f(t) \sin n\omega_0 t\, dt$
Odd n only	0	$\dfrac{4}{T}\displaystyle\int_0^{T/2} f(t) \cos n\omega_0 t\, dt$	$\dfrac{4}{T}\displaystyle\int_0^{T/2} f(t) \sin n\omega_0 t\, dt$

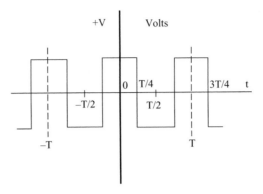

FIGURE 12.6: Square waveform for example problem

By inspection, the average value over one period is zero so that $a_0 = 0$.

Value of a_1 with $n = 1$ is

$$a_n = \frac{2}{T}\left(V\int_0^{T/4}\cos n\omega_0 t\, dt - V\int_{T/4}^{3/4\,T}\cos n\omega_0 t\, dt + V\int_{3/4\,T}^{T}\cos n\omega_0 t\, dt\right)$$

$$a_n = \frac{2V}{T}\left[\left(\frac{1}{n\omega_0}\sin n\omega_0 t\right)\Big|_0^{T/4} - \left(\frac{1}{n\omega_0}\sin n\omega_0 t\right)\Big|_{T/4}^{3/4\,T} + \left(\frac{1}{n\omega_0}\sin n\omega_0 t\right)\Big|_{3/4\,T}^{T}\right]$$

$$a_n = \frac{2V}{\omega_0 T}\left[\left(\sin\frac{\omega_0 T}{4}\right) - \left(\sin\frac{3\omega_0 T}{4} - \sin\frac{\omega_0 T}{4}\right) + \left(\sin\omega_0 T - \sin\frac{3\omega_0 T}{4}\right)\right]$$

Since $\omega_0 T = \frac{2\pi}{T}\cdot T = 2\pi$, then

for $n = 1$, $a_1 = \frac{2V}{2\pi}\left[\sin\frac{\pi}{2} - \left(\sin\frac{3\pi}{2} - \sin\frac{\pi}{2}\right) + \left(\sin 2\pi - \sin\frac{3\pi}{2}\right)\right]$

$$a_1 = \frac{V}{\pi}[1 - (-1 - 1) + (0 - (-1))] = \frac{V}{\pi}[1 + 2 + 1] = \frac{4V}{\pi}$$

$$a_n = \frac{2}{T}\left(V\int_0^{T/4}\cos n\omega_0 t\, dt - V\int_{T/4}^{3/4\,T}\cos n\omega_0 t\, dt + V\int_{3/4\,T}^{T}\cos n\omega_0 t\, dt\right)$$

$$a_n = \frac{2V}{T}\left[\left(\frac{1}{n\omega_0}\sin n\omega_0 t\right)\Big|_0^{T/4} - \left(\frac{1}{n\omega_0}\sin n\omega_0 t\right)\Big|_{T/4}^{3/4\,T} + \left(\frac{1}{n\omega_0}\sin n\omega_0 t\right)\Big|_{3/4\,T}^{T}\right]$$

Since $2\omega_0 T = 4\pi$, then:

For $n = 2$, $a_2 = \dfrac{2V}{2\omega_0 T}\left[\sin 2\omega_0 \dfrac{T}{4} - \left(\sin 2\omega_0 \dfrac{3T}{4} - \sin 2\omega_0 \dfrac{T}{4}\right)\left(\sin 2\omega_0 T - \sin 2\omega_0 \dfrac{3T}{4}\right)\right]$

$a_2 = \dfrac{2V}{4\pi}\left[\sin \dfrac{4\pi}{4} - \left(\sin \dfrac{4\pi \cdot 3}{4} - \sin \pi\right) + (\sin 4\pi - \sin 3\pi)\right]$

$a_2 = \dfrac{2V}{4\pi}[0 - (0 - 0) + (0 - 0)] = 0$

Likewise, $a_n = 0$ for all even values of a_n.

Since $\omega_0 T = 2\pi$, then $n\omega_0 T$ for $n = 3$ is $3\omega_0 T = 6\pi$.

For $n = 3$, $a_3 = \dfrac{2V}{6\pi}\left[\sin \dfrac{6\pi}{4} - \left(\sin \dfrac{6\pi \cdot 3}{4} - \sin \dfrac{6\pi}{4}\right) + \left(\sin 6\pi - \sin \dfrac{6\pi \cdot 3}{4}\right)\right]$

$a_3 = \dfrac{V}{3\pi}[-1 - (+1 - (-1)) + 0 - (+1)]$

$a_3 = \dfrac{V}{3\pi}[-1 - 2 - 1] = -\dfrac{4V}{3\pi}$

Applying the same procedure for all n, we find that

$$a_n = \begin{cases} +\dfrac{4V}{n\pi}, & n = 1, 5, 9, \ldots \\ -\dfrac{4V}{n\pi}, & n = 3, 7, 11, \ldots \\ 0, & n = \text{even} \end{cases}$$

The imaginary term b_n values are calculated as follows:

$b_n = \dfrac{2}{T}\left(V\int_0^{T/4} \sin n\omega_0 t\, dt - V\int_{T/4}^{3/4\,T} \sin n\omega_0 t\, dt + V\int_{3/4\,T}^{T} \sin n\omega_0 t\, dt\right)$

$b_n = \dfrac{2V}{n\omega_0 T}\left[(-\cos n\omega_0 t)\Big|_0^{T/4} - (-\cos n\omega_0 t)\Big|_{T/4}^{3/4\,T} + (-\cos n\omega_0 t)\Big|_{3/4\,T}^{T}\right]$

$b_1 = \dfrac{2V}{n\omega_0 T}\left[-\cos \dfrac{2\pi}{4} + \cos 0 - \left(-\cos \dfrac{6\pi}{4} + \cos \dfrac{2\pi}{4}\right) + \left(-\cos 2\pi + \cos \dfrac{6\pi}{4}\right)\right]$

$b_1 = \dfrac{2V}{n\omega_0 T}[0 + 1 + 0 - 0 - 1 + 0] = 0$

$b_1 = \dfrac{2V}{2\pi}(0) = 0$

If you continue this process for other integers of n, you will obtain $b_n = 0$ for all values of n.

Before moving into Discrete and Fast Fourier transformations and frequency-domain analysis, let me regress into perhaps the most important function to engineers, physicists, and mathematicians—the **exponential function**—which has the property that the derivative and integral yield a function proportional to itself as in (12.10).

$$\frac{d}{dt}e^{st} = se^{st} \quad \text{and} \quad \int e^{st}\,dt = \frac{1}{s}e^{st} \tag{12.10}$$

It turns out that every function or waveform encountered in practice can always be expressed as a sum of various exponential functions.

12.5   EULER EXPANSION

Euler showed that the expression, $s = -j\omega$, of a complex frequency represents signals oscillating at angular frequency, ω, and that the $\cos \omega t$ could thereby then be represented as (12.11).

$$\cos(\omega t) = \frac{e^{j\omega t} + e^{-j\omega t}}{2} \tag{12.11}$$

In a way, the Discrete Fourier Series is a method of representing periodic signals by exponentials whose frequencies lie along the $j\omega$ axis. Expressing the Fourier Trigonometric Series in an equivalent form in terms of exponentials, or $e^{\pm jn\omega t}$:

$$f(t) = a_0 + \sum_{n=1}^{\infty}(a_n \cos n\omega_0 t + b_n \sin n\omega_0 t)$$

$$\cos n\omega_0 t = \frac{1}{2}\left(e^{jn\omega_0 t} + e^{-jn\omega_0 t}\right)$$

$$\sin n\omega_0 t = \frac{1}{2j}\left(e^{jn\omega_0 t} - e^{-jn\omega_0 t}\right)$$

$$\text{then } f(t) = a_0 + \sum_{n=1}^{\infty}\left(a_n\frac{e^{jn\omega_0 t} + e^{-jn\omega_0 t}}{2} + b_n\frac{e^{jn\omega_0 t} - e^{-jn\omega_0 t}}{2j}\right)$$

Grouping terms, $-j = \frac{1}{j}$:

$$f(t) = a_0 + \sum_{n=1}^{\infty}\left(\left(\frac{a_n - jb_n}{2}\right)e^{jn\omega_0 t} + \left(\frac{a_n + jb_n}{2}\right)e^{-jn\omega_0 t}\right)$$

Redefining coefficients:

$$\tilde{c}_n = \frac{a_n - jb_n}{2}, \quad \tilde{c}_{-n} = \frac{a_n + jb_n}{2}, \quad \text{and} \quad \tilde{c}_0 = a_0$$

Then, the Fourier expression for a function may be written as (12.12):

$$f(t) = \tilde{c}_0 + \sum_{n=1}^{\infty} \left(\tilde{c}_n e^{jn\omega_0 t} + \tilde{c}_{-n} e^{-jn\omega_0 t} \right) \tag{12.12}$$

Letting n range from 1 to ∞ is equivalent to letting n range from $-\infty$ to $+\infty$, including 0.

$$\text{Then } f(t) = \sum_{n=-\infty}^{\infty} \tilde{c}_n e^{jn\omega_0 t}$$

The coefficients $\tilde{c}_n$ can be evaluated in terms of a_n and b_n as shown in (12.13).

$$|\tilde{c}_n| = \frac{1}{2}\sqrt{a_n^2 + b_n^2} = \frac{1}{2}c_n \quad \text{and} \quad \phi_n = \tan^{-1}\frac{b_n}{a_n} \tag{12.13}$$

$$\tilde{c}_n = |\tilde{c}_n| \ e^{j\theta n}$$

$$\tilde{c}_n = \tilde{c}_n^* = |\tilde{c}_n| \ e^{-j\theta n}$$

Note that in (12.13), the coefficients of the exponential Fourier Series are $^1/_2$ the magnitude of the geometric Fourier Series coefficients.

12.6   LIMITATIONS

Overall, the Fourier Series method has several limitations in analyzing linear systems for the following reasons:

1. The Fourier Series can be used only for inputs that are periodic. Most inputs in practice are nonperiodic.

2. The method applies only to systems that are stable (systems whose natural response decays in time).

The first limitation can be overcome if we can represent the nonperiodic input, $f(t)$, in terms of exponential components, which can be done by either the Laplace or

Fourier Transform representation of $f(t)$. Consider the following nonperiodic signal, $f(t)$, which one would like to represent by eternal exponential functions. As a result, we can construct a new periodic function, $f_T(t)$, with period T, where the function $f(t)$ is represented every T seconds. The period, T, must be large enough so there is not any overlap between pulse shapes of $f(t)$. The new function is a periodic function and can be represented with an exponential Fourier Series.

12.7   LIMITING PROCESS

What happens if the interval, T, becomes infinite for a pulse function, $f(t)$? Do the pulses repeat after an infinite interval? Such that $f_T(t)$ and $f(t)$ are identical in the limit, and the Fourier Series representing the periodic function, $f_T(t)$, will also represent $f(t)$?

Discussion on Limiting Process: Let the interval between pulses become infinite, $T = \infty$, in the series, so we can represent the exponential Fourier Series for $f_T(t)$ as in (12.14).

$$\sum_{n=-\infty}^{\infty} F_n \, e^{jn\omega_0 t} \tag{12.14}$$

Where ω_0 is the fundamental frequency and F_n is the term which represents the amplitude of the component of frequency, $n\omega_0$, the coefficient. As T becomes larger, the fundamental frequency, ω_0, becomes smaller, and the amplitude of individual components also become smaller as shown in (12.15). The frequency spectrum becomes denser, but its shape is unaltered.

$$\text{Limit } T \to \infty, \omega_0 \to 0 \tag{12.15}$$

The spectrum exists for every value of ω and is no longer discrete but a continuous function of ω. So, let us denote the fundamental frequency ω_0 as being infinite. The logic is as follows:

$$\text{Since Limit } T \to \infty, \omega_o \to 0, \text{ then } T = \frac{2\pi}{\Delta\omega},$$
$$\text{and } TF_n \text{ is a function of } jn\Delta\omega, \text{ or } TF = F\,(jn\Delta\omega).$$

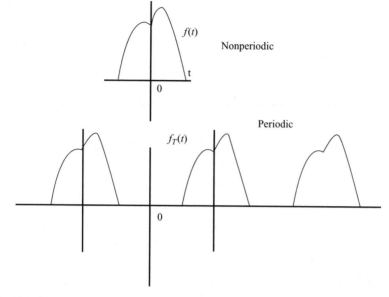

FIGURE 12.7: A transient pulse. Top trace is a transient; bottom trace is periodic

It is an accepted fact that a periodic function, $f_T(t)$, can be expressed as the sum of eternal exponentials of frequencies $0, \pm j\Delta\omega, \pm j2\Delta\omega, \pm 3j\Delta\omega \ldots$ etc. Also, it is known that the Fourier Transform, F_n, amplitude of the component of frequency, $jn\,\Delta\omega$, is expressed as $F(jn\Delta\omega)\,{}^{\Delta\omega}/_{2\pi}$.

Proof of the Limiting Process: Viewing Fig. 12.7, let us examine the proof of the limiting process.

Let us begin with the general expression (12.16):

$$\text{In the Limit as } T \to \infty : \lim_{T\to\infty} f_T(t) = f(t) \qquad (12.16)$$

Then the exponential Fourier Series for $f_T(t)$ is given by (12.17).

$$f_T(t) = \sum_{n=-\infty}^{\infty} F_n\, e^{jn\omega_0 t} \qquad (12.17)$$

where $\omega_0 = \frac{2\pi}{T}$ and $F_n = \frac{1}{T}\int_{-T/2}^{T/2} f_T(t)\, e^{-jn\omega_0 t}\, dt$

Since Limit $T \to \infty$, $\omega_0 \to 0$, ω_0 can be denoted by $\Delta\omega$ (see (12.18))

Then $\qquad T = \dfrac{2\pi}{\omega_0} = \dfrac{2\pi}{\Delta\omega}$ and $TF_n = \dfrac{1}{T}\displaystyle\int_{-T/2}^{T/2} f_T(t)\, e^{-jn\omega_0 t}\, dt$ $\qquad$ (12.18)

Since $TF_n(jn\Delta\omega)$, the expression may be rewritten as $TF_n = F(jn\Delta\omega)$.

Then (12.17), $f_T(t) = \displaystyle\sum_{n=-\infty}^{\infty} F_n e^{jn\omega_0 t}$, becomes (12.19).

$$f_T(t) = \sum_{n=-\infty}^{\infty} \frac{F(jn\Delta\omega)}{T}\, e^{(jn\omega_0 t)t}$$

$$f_T(t) = \sum_{n=-\infty}^{\infty} \left[\frac{F(jn\Delta\omega)}{2\pi} \Delta\omega \right] e^{(jn\omega_0 t)t} \qquad (12.19)$$

Rewriting (12.19) in the "Limit": as $T \to \infty$, $\omega_0 \to 0$, $f_T(t) \to f(t)$, the equation takes the form of (12.20), thus proving the limit process.

$$f(t) = \operatorname*{Lim}_{T\to\infty} f_T(t) = \left[\operatorname*{Lim}_{\Delta\omega\to 0} \frac{1}{2\pi} \sum F(jn\Delta\omega) e^{jn\Delta\omega t} \Delta\omega \right] \qquad (12.20)$$

12.8   INVERSE FOURIER TRANSFORM

To return to the "Time Domain" after transforming into the "Frequency Domain," the inverse Fourier transform must be obtained. By definition, the Inverse Fourier Transform is written as (12.21). The process is the same as taking the forward transform, except that in lieu of multiplying the time domain function, $f(t)$, by the Fourier Basis Function, all the Fourier Coefficients form the Frequency domain function, $F_n = F(jn\Delta\omega)\, {}^{\Delta\omega}\!/_{2\pi}$, are multiplied by the inverse Fourier basis function as in (12.19). Particular emphasis is made to point out that the sign of the basis function exponential frequency, $j\omega t$, is positive for the inverse transformation.

$$f(t) = \frac{1}{2\pi} \int_{-\infty}^{\infty} F(j\omega)\, e^{j\omega t}\, d\omega \qquad (12.21)$$

where $n\Delta\omega$ becomes ω (a continuous variable) and the function $F(j\omega) = \operatorname*{Lim}_{\Delta\omega\to 0} F(jn\Delta\omega)$.

12.9 SUMMARY

The "Forward Fourier Transform" equations are summarized in (12.22) and (12.23).

$$F_n = \frac{1}{T} \int_{-T/2}^{T/2} f_T(t)\, e^{-jn\omega_0 t}\, dt = \frac{F(jn\Delta\omega)^*}{T} \tag{12.22}$$

then $F(j\omega) = \underset{T\to\infty}{Lim} \int_{-T/2}^{T/2} f_T(t)\, e^{-jn\Delta\omega t}\, dt$ frequency spectrum of $f_T(t)$

$$\text{or for the infinite case: } F(j\omega) = \int_{-\infty}^{\infty} f(t)\, e^{-j\omega t}\, dt \tag{12.23}$$

where in the Limit as $T \to \infty$, $\Delta\omega \to 0$, $f_T(t) \to f(t)$.

The Discrete Inverse Fourier Transform $\underset{\Delta\omega\to 0}{Lim} \frac{1}{2\pi} \sum F(j\omega)\, e^{j\Delta\omega t} \Delta\omega$ is, by definition, the integral, which may be expressed as (12.24).

$$f(t) = \frac{1}{2\pi} \int_{-\infty}^{\infty} F(j\omega)\, e^{j\omega t}\, d\omega \tag{12.24}$$

where $n\Delta\omega$ becomes a continuous variable, ω, and the function $F(j\omega)$ is $F(jn\Delta\omega)$.

The result is that $F(j\omega) = \int_{-\infty}^{\infty} f(t)\, e^{-j\omega t}\, dt$ is the presentation of the nonperiodic function, $f(t)$, in terms of exponential functions. The amplitude of the component of any frequency, ω, is proportional to $F(j\omega)$. Therefore, $F(j\omega)$, represents the frequency spectrum of $f(t)$ and is called the *Frequency Spectral of a Function*. Equations (12.23) and (12.24) are referred to as the Fourier Transform pair. Equation (12.23) is the *Direct Fourier Transform* of $f(t)$, and (12.24) is the *Inverse Fourier Transform*.

CHAPTER 13

Fast Fourier Transform

The "Fast Fourier Transform," referred to as the "FFT," is a computational tool that facilitates signal analysis, such as Power Spectral Analysis, by means of digital computers. The FFT is a method for efficiently computing the Discrete Fourier Transform (DFT) of a digitized time series or discrete data samples. The FFT takes advantage of the symmetrical properties of periodic sinusoidal waveforms. Before development of the FFT, transformation of signals into the frequency domain was done by the standard trigonometric Fourier Series computational procedures.

13.1 CONTINUOUS FOURIER TRANSFORM

Before starting on the FFT, let us review or recall from Chapter 12 that the frequency spectrum of an analog signal $x(t)$ may be obtained by taking the Fourier Transform as shown in (13.1):

$$F(\omega) = \int_{-\infty}^{\infty} x(t)e^{-j\omega t} dt \qquad (13.1)$$

where $x(t)$ and $F(\omega)$ are complex functions.

13.2 DISCRETE FOURIER TRANSFORM

Also recall that the sampled signal may be written either $x(nT)$ or $x(n)$,

$$x(t) \xrightarrow[\text{Interval, } T]{\text{Sampling}} x(nT) \text{ or } x(n)$$

where T-sampling period and n-number of samples, equally spaced with T as the sampling interval.

Then the Discrete Fourier Transform equation is given by (13.2):

$$x(\omega) = \sum_{n=-\infty}^{\infty} x(n)e^{-j\omega n} \qquad (13.2)$$

where ω ranges from 0 to 2π and $x(\omega)$ is periodic with 2π

13.3   DEFINITION OF SAMPLING RATE (OR SAMPLING FREQUENCY)

Let us review highlights of Chapter 6 on Sampling Theory before discussing the Fast Fourier Transform (FFT). The basis of the Nyquist sampling theorem is the premise that in order to successfully represent and/or recover an original analog signal, the analog signal must be sampled at least twice the highest frequency in the signal. If the highest frequency in the analog signal is known, then it can be sampled at twice the highest frequency, f_n. In terms of time, the spacing between samples is called time period or sampling interval, T. Thus,

$$T \leq 1(/2f_n) \quad \text{or} \quad f_{\text{sampling}} \geq 2f_n = 1/T$$

where T is the sampling period (or sampling interval), and f_{sampling} is the sampling frequency or sampling rate.

It should be noted that the maximum frequency, f_n, is also called the *Cutoff Frequency*, the *Nyquist Frequency*, or the *Folding Frequency*.

If the highest frequency in the signal is not known, then the signal should be band-limited with an analog low-pass filter, and then the filtered signal should be sampled at least twice the upper frequency limit of the filter bandwidth. If the low-pass filtering procedure is not followed, then the problem of "Aliasing" arises. The Aliasing problem is usually caused when the spacing between samples is very large, which corresponds to sampling at a rate less than twice the highest frequency in the signal.

Aliasing is often defined as "the contamination of low frequencies by unwanted high frequency signals/noise" (Fig. 13.1), which by itself is an incorrect statement without

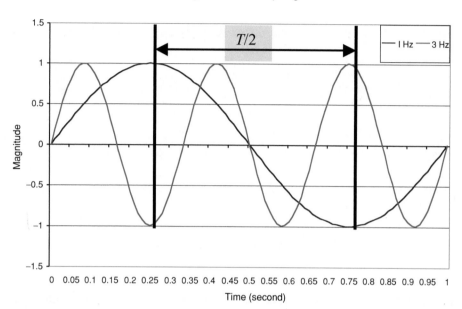

FIGURE 13.1: Aliasing. The example shows that a 3-Hz waveform has the same half period ($T/2$) as a 1-Hz waveform, if the signals are sampled at 2 samples per second

some clarification. Aliasing can be eliminated by proper sampling rate (meaning at least twice higher than the highest frequency in the signal) or by low-pass filtering before sampling the signal at a rate not less than twice the known low-pass filter cutoff frequency if the highest frequency in the signal is not known. What is implied by the statement is that if the period of the sampling window is equal to, or less than, one over twice the highest frequency in the signal, $T \leq 1/(2f_n)$, then in the frequency domain, the spectrum is repeated every $2\pi * f_n$ Hz. In general, the spectrum repeats over every 2π range from $-\infty$ to $+\infty$. Thus, it would be sufficient to obtain values between 0 and 2π in a DFT.

13.3.1 Short-Hand Notation of the Discrete Fourier Transform (DFT)

The DFT is often written as in (13.3).

$$x(k) = \sum_{n=0}^{N-1} x(n) W_N^{nk} \quad \text{for} \quad k = 0, 1, \ldots, N-1 \tag{13.3}$$

TABLE 13.1: Spectral Window vs. Frequency Resolution

	$f_{\text{sampling rate}}$	N	ΔT
1.	1 sec − 2000 samples/sec	2000	0.0005
2.	1 sec − 1000 samples/sec	1000	0.001
3.	2 sec − 1000 samples/sec	2000	0.05

Since $w_n = 2\pi/N = 2\pi\, T$. The frequency resolutions are shown below.

1.	0.003141592	$f_{\text{res}} = \dfrac{\omega_0}{2\pi\mathrm{T}}$	=1.0 Hz
2.	0.006283185	$f_{\text{res}} = \dfrac{\omega_n{}^* f_{\text{samp}}}{2\pi}$	=1.0 Hz
3.	0.00314592		=0.5 Hz

where $W_N = e^{-j2\pi/N}$ is called phase or *The Twiddle Factor*, $x(k)$ is called the N-point DFT, and the distance between successive samples in the frequency domain gives the fundamental frequency of $x(t)$.

In terms of normalized frequency units, the fundamental frequency is given by $2\pi/N$. Note in the example given in Table 13.1 that a one-second window results in a 1-Hz frequency resolution regardless of the sampling rate.

By definition, the Frequency Resolution (f_{res}) or the fundamental frequency of the spectral (f_0) is "The minimum frequency that a signal can be resolved in the frequency domain, and is equal to the inverse of the record length or the signal window length being analyzed (in seconds) in time domain" (13.4):

$$f_{\text{res}} = \frac{1}{R.L} = \frac{1}{N\mathrm{T}} \tag{13.4}$$

Record length: The length of the sampled analog signal with a total of N samples is given as $R.L = NT$.

Expanding the DFT equation results in (13.5):

$$x(k) = \sum_{n=0N}^{N-1} \{(\mathrm{Re}[x(n)]\mathrm{Re}[W_N^{nk}] - \mathrm{Im}[x(n)]\mathrm{Im}[W_N^{nk}]) \\ + (\mathrm{Re}[x(n)]\mathrm{Im}[W_N^{nk}] + \mathrm{Im}[x(n)]\mathrm{Re}[W_N^{nk}])\} \tag{13.5}$$

where $k = 0, 1, \ldots, N-1$

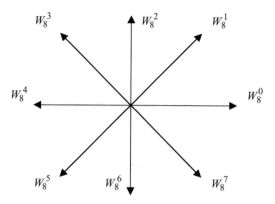

FIGURE 13.2: Symmetry and periodicity of W_N for $N = 8$

For each k, (13.5) needs $4N$ real multiplications and $(4N - 2)$ real additions. The computation of $x(k)$ needs $4N^2$ real multiplications and $N(4N - 2)$ real additions. In terms of complex operations, there are N^2 complex multiplication and $N(N - 1)$ complex additions. As N increases, it can be seen that the number of computations become tremendously large. Hence, there was a need to improve the efficiency of DFT computation and to exploit the symmetry and periodicity properties of the twiddle factor, W_N, as shown in (13.6) and Fig. 13.2.

$$\text{Symmetry}: W_N^k = -W_N^{k+\frac{N}{2}}$$
$$\text{Periodicity}: W_N^k = W_N^{N+k}$$

(13.6)

13.4 COOLEY-TUKEY FFT (DECIMIMATION IN TIME)

The Cooley–Tukey FFT was one of the first FFTs developed. Suppose a time series having N samples is divided into two functions, of which each function has only half of the data points ($N/2$). One function consists of the even numbered data points (x_0, x_2, x_4 ..., etc,), and the other function consists of the odd numbered data points (x_1, x_3, x_5, ..., etc.). Those functions may be written as

$$Y_k = x_{2k}; \quad Z_k = x_{2k+l}, \quad \textit{for:} \quad k = 0, 1, 2, \ldots, (N/2) - 1$$

Since Y_k and Z_k are sequences of $N/2$ points each, the two sequences have Discrete Fourier transforms, which are written in terms of the odd and even numbered data points. For values of r greater than $N/2$, the discrete Fourier transforms Br and Cr repeat periodically the values taken on when $r < N/2$. Therefore, substituting $r + N/2$ for r in (13.7), one can obtain (13.8) by using the Twiddle Factor, $W = \exp(-2\pi j/N)$.

$$A_r = B_r + W^r C_r \tag{13.7}$$

$$A_{r+\frac{N}{2}} = B_r - W^r C_r \tag{13.8}$$

From (13.7) and (13.8), the first $N/2$ and last $N/2$ points of the discrete Fourier transform of the data x_k can be simply obtained from the DFT of both sequences of $N/2$ samples. Assuming that one has a method that computes DFTs in a time proportional to the square of the number of samples, the algorithm can be used to compute the transforms of the odd and even data sequences using the Ar (13.7) and $Ar + N/2$ (13.8) equations to find Ar. The N operations require a time proportional to $2(N/2)^2$.

Since it has been shown that the computation of the DFT of N samples can be reduced to computing the DFTs of two sequences of $N/2$ samples, the computation of the sequence B_k or C_k can be reduced by decimation in time to the computation of sequences of $N/4$ samples. These reductions can be carried out as long as each function has a number of samples that are divisible by 2. Thus, if $N = 2^n$, n such reductions (often referred to as "Stages") can be made.

For example, the successive decimation (reduction) of a Discrete Fourier Transform will result in a Fast Fourier Transform with fewer computations. In general, $N \log_2 N$ complex additions and at most $(1/2)N\log_2 N$ complex multiplications are required for computation of the Fast Fourier Transform of an N point sequence, where N is a power of 2 ($N = 2^n$).

13.4.1 Derivation of FFT Algorithm

Consider the N-point DFT that can be reduced into two $N/2$ DFTs (even and odd indexed sequences), as $x(n)$ is decomposed into two $N/2$-point sequences (decimation-in-time) and is shown in (13.9).

$$\therefore x(k) = \sum_{n=0}^{\frac{N}{2}-1} x(2n) W_N^{2nk} + \sum_{n=0}^{\frac{N}{2}-1} x(2n+1) W_N^{(2n+1)k} \tag{13.9}$$

where: $W_N^2 = \left[e^{-j\frac{2\pi}{N}} \right]^2 = e^{-j\frac{2\pi}{(N/2)}} = W_{\frac{N}{2}}$

Then, (13.9) can then be rewritten as (13.10).

$$\therefore x(k) = \sum_{n=0}^{\frac{N}{2}-1} x(2n) W_{\frac{N}{2}}^{nk} + W_N^k \sum_{n=0}^{\frac{N}{2}-1} x(2n+1) W_{\frac{N}{2}}^{nk} \qquad (13.10)$$

where $x_1(k)$ is $= x_1(k) + W_N^k x_2(k)$, $k = 0, 1 \ldots, N$.

The sequences, $x_1(k)$ and $x_2(k)$ can each be further divided into even and odd sequences as shown in (13.11) and (13.12).

$$x_1(k) = \sum_{n=0}^{\frac{N}{2}-1} x(2n) W_{N/2}^{nk}, \quad \text{for} \quad k = 0, 1, \ldots, \frac{N}{2} - 1 \qquad (13.11)$$

$$= \sum_{n=0}^{\frac{N}{4}-1} x(4n) W_{N/4}^{nk} + W_{N/2}^k \sum_{n=0}^{\frac{N}{4}-1} x(4n+2) W_{N/4}^{nk}, \quad \text{for} \quad k = 0, 1, \ldots, \frac{N}{4} - 1 \qquad (13.12)$$

and for the $x_2(k)$ sequence as shown in (13.13) and (13.14):

$$x_2(k) = \sum_{n=0}^{\frac{N}{2}-1} x(2n+1) W_{N/2}^{nk}, \quad \text{for} \quad k = 0, 1, \ldots, \frac{N}{2} - 1 \qquad (13.13)$$

$$= \sum_{n=0}^{\frac{N}{4}-1} x(4n+1) W_{N/4}^{nk} + W_{N/2}^k \sum_{n=0}^{\frac{N}{4}-1} x(4n+3) W_{N/4}^{nk}, \quad \text{for} \quad k = 0, 1, \ldots, \frac{N}{4} - 1 \qquad (13.14)$$

Table 13.2 shows the reduction in computations between the DFT and the FFT.

TABLE 13.2: Comparison DFT and FFT Computation with $N = 8$		
COMPLEX	**DFT**	**FFT**
Multiplications	$N^2 = 64$	$N \log_2 N = 24$
Additions	$N(N-1) = 56$	$N \log_2 1 = 24$
For $N = 65\,536$		
Complex multiplications	4.3×10^9	1.049×10^6

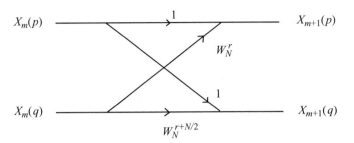

FIGURE 13.3: FFT "Butterfly" signal flow diagram

13.5   THE FFT "BUTTERFLY" SIGNAL FLOW DIAGRAM

To predict the next value in the stage (decimation), the previous stage values are used as inputs, where p may indicate the odd sequence and q the even sequence. The flow graph (Fig. 13.3) is called the "butterfly computation" because of its appearance.

The output of the stage is calculated with (13.15) and (13.16).

$$X_{m+1}(p) = X_m(p) + W_N^r X_m(q) \tag{13.15}$$

$$X_{m+1}(q) = X_m(p) + W_N^{r+\frac{N}{2}} X_m(q) \tag{13.16}$$

However, from symmetry and periodicity properties:

$$W_N^{\frac{N}{2}} = e^{-j\left(\frac{2\pi}{N}\right)\left(\frac{N}{2}\right)} = e^{-j\pi} = -1$$

the reduced Butterfly flow diagram becomes Fig. 13.4 and the equations are rewritten as

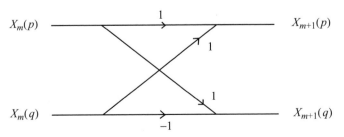

FIGURE 13.4: Reduced Butterfly flow diagram

(13.17) and (13.18).

$$X_{m+1}(p) = X_m(p) + W_N^r X_m(q) \qquad (13.17)$$

$$X_{m+1}(q) = X_m(p) - W_N^r X_m(q) \qquad (13.18)$$

Bit Reversal : To perform the computation as shown in the Fig. 13.4 flow graph, the input data must be stored in a nonsequential order and is known as the bit-reverse order. For a sequence of $N = 8$, the input data are represented in binary form as

LSB $= 0$

$\quad X(0) = X(0, 0, 0)$

$\quad X(4) = X(1, 0, 0)$

$\quad X(2) = X(0, 1, 0)$

$\quad X(6) = X(1, 1, 0)$

LSB $= 1$

$\quad X(1) = X(0, 0, 1)$

$\quad X(5) = X(1, 0, 1)$

$\quad X(3) = X(0, 1, 1)$

$\quad X(7) = X(1, 1, 1)$

Upon observation, one can see that the bits (LSB $= 0$) of the top section are all even number samples; and LSB $= 1$ of bottom section are all odd number samples, which are obtained by writing the signal sequences in binary form and bit-reversing each to get the new order. If the input is not bit-reversed, then the output will need to be bit-reversed. The flow graph can be arranged in such a way that both input and output need not be bit-reversed.

13.6   DECIMATION-IN-FREQUENCY

The configuration is obtained by dividing the output sequence into smaller and smaller subsequences. Here the input sequence is divided into first or second halves. According to Fourier integral theorem, a function must satisfy the Dirichlet conditions or every finite interval and the integral from infinity to negative infinity is finite. The conditions are

1) the function should have a finite number of maxima or minima,

2) the function should have a finite number of discontinuities, and

3) the function should have a finite value or $\int_{t}^{t-T} f(t)\, dt$ is less than infinity.

Conditions for the Fourier Series include the following:

1) periodic functions (practical is nonperiodic);

2) can only be applied to stable systems (systems where natural response decays in time, for example, convergent).

Example Problem: Consider the sinusoidal waveform of Fig. 13.5.

According to Nyquist sampling theorem, the waveform in Fig. 13.5 should be sampled at more than two samples per second to successfully reconstruct the signal. For the sake of convenience, consider eight samples per second of the above waveform. Therefore: $N = 8 = 2^n = 2^3$, N is the total number of samples, and $n = 3$ stages; $T = 125$ milliseconds, sampling interval; Record length $= 1$ second (time of recording data $= N \times T$) Frequency resolution $= 1$ Hz $(1/(N \times T))$

The discrete signal is given by (13.19).

$$X(nT) = \cos\left(2\pi f n T\right) = \cos\left(\frac{2\pi f n}{N}\right) \tag{13.19}$$

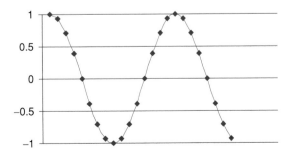

FIGURE 13.5: Trace of the waveform is $x(t) = \cos(\omega t)$, where $\omega = 2\pi f$ and $f = 1$ Hz

Symmetric property of twiddle factor implies:

$$W_N^k = W_N^{k+\frac{N}{2}} = W_8^k = -W_8^{k+4}, \text{ which gives}$$
$$W_8^0 = -W_8^4 \qquad W_8^2 = -W_8^6$$
$$W_8^1 = -W_8^5 \qquad W_8^3 = -W_8^7$$

13.6.1 Butterfly First-Stage Calculations

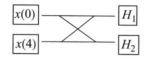

$$H_1 = x(0) + W_8^0 x(4) = (1 + j \times 0) + (1 + j \times 0)(-1 + j \times 0) = 0 + j \times 0$$
$$H_2 = x(0) - W_8^0 x(4) = (1 + j \times 0) - (1 + j \times 0)(-1 + j \times 0) = 2 + j \times 0$$

$$H_3 = x(2) + W_8^0 x(6) = (0 + j \times 0) + (1 + j \times 0)(0 + j \times 0) = 0 + j \times 0$$
$$H_4 = x(2) - W_8^0 x(6) = (0 + j \times 0) - (1 + j \times 0)(0 + j \times 0) = 0 + j \times 0$$

$$H_5 = x(1) + W_8^0 x(5) = (0.707 + j \times 0) + (1 + j \times 0)(-0.707 + j \times 0) = 0 + j \times 0$$
$$H_6 = x(1) - W_8^0 x(5) = (0.707 + j \times 0) - (1 + j \times 0)(-0.707 + j \times 0)$$
$$= 1.414 + j \times 0$$

$$H_7 = x(3) + W_8^0 x(7) = (-0.707 + j \times 0) + (1 + j \times 0)(0.707 + j \times 0) = 0 + j \times 0$$
$$H_8 = x(3) - W_8^0 x(7) = (-0.707 + j \times 0) - (1 + j \times 0)(0.707 + j \times 0)$$
$$= -1.414 + j \times 0$$

It should be noted that only $W_8{}^0 \rightarrow W_8{}^0$ needs to be calculated for the first stage.

13.6.2 Second-Stage Calculations

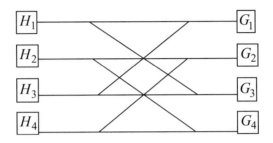

$$G_1 = H_1 + W_8^0 H_3 = (0 + j \times 0) + (1 + j \times 0)(0 + j \times 0) = 0 + j \times 0$$

$$G_2 = H_2 + W_8^0 H_4 = (2 + j \times 0) + (1 + j \times 0)(0 + j \times 0) = 2 + j \times 0$$

$$G_3 = H_1 - W_8^0 H_3 = (0 + j \times 0) - (1 + j \times 0)(0 + j \times 0) = 0 + j \times 0$$

$$G_4 = H_2 - W_8^0 H_4 = (2 + j \times 0) - (1 + j \times 0)(0 + j \times 0) = 2 + j \times 0$$

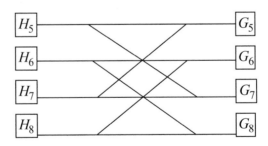

$$G_5 = H_5 + W_8^0 H_7 = (0 + j \times 0) + (1 + j \times 0)(0 + j \times 0) = 0 + j \times 0$$

$$G_6 = H_6 + W_8^0 H_8 = (1.414 + j \times 0) + (1 + j \times 0)(-1.414 + j \times 0)$$

$$= 1.414 - j \times 1.414$$

$$G_7 = H_5 - W_8^0 H_7 = (0 + j \times 0) - (1 + j \times 0)(0 + j \times 0) = 0 + j \times 0$$

$$G_8 = H_6 - W_8^0 H_8 = (1.414 + j \times 0) - (1 + j \times 0)(-1.414 + j \times 0)$$

$$= 1.414 + j \times 1.414$$

13.6.3 Third-Stage Calculations

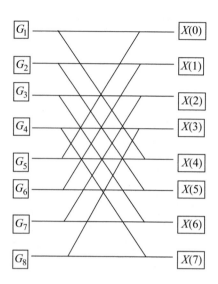

$$X(0) = G_1 + W_8^0 G_5 = (0 + j \times 0) + (1 + j \times 0)(0 + j \times 0) = 0 + j \times 0$$

$$X(1) = G_2 + W_8^1 G_6 = (2 + j \times 0) + (0.707 + j \times 0.707)(1.414 - j \times 1.414)$$
$$= 4 + j \times 0$$

$$X(2) = G_3 + W_8^2 G_7 = (0 + j \times 0) + (0 + j \times 1)(0 + j \times 0) = 0 + j \times 0$$

$$X(3) = G_4 + W_8^3 G_8 = (2 + j \times 0) + (-0.707 + j \times 0.707)(1.414 + j \times 1.414)$$
$$= 0 + j \times 0$$

$$X(4) = G_1 - W_8^0 G_5 = (0 + j \times 0) - (1 + j \times 0)(0 + j \times 0) = 0 + j \times 0$$

$$X(5) = G_2 - W_8^1 G_6 = (2 + j \times 0) - (0.707 + j \times 0.707)(1.414 - j \times 1.414)$$
$$= 0 + j \times 0$$

$$X(6) = G_3 - W_8^2 G_7 = (0 + j \times 0) - (0 + j \times 1)(0 + j \times 0) = 0 + j \times 0$$

$$X(7) = G_4 - W_8^3 G_8 = (2 + j \times 0) - (-0.707 + j \times 0.707)(1.414 + j \times 1.414)$$
$$= 4 + j \times 0$$

Each of the eight samples for $x(nt)$ is given by

$$x(0) = 1 + j \times 0 \qquad\qquad x(4) = -1 + j \times 0$$
$$x(1) = 0.707 + j \times 0 \qquad x(5) = -0.707 + j \times 0$$
$$x(2) = 0 + j \times 0 \qquad\qquad x(6) = 0 + j \times 0$$
$$x(3) = -0.707 + j \times 0 \qquad x(7) = 0.707 + j \times 0$$

13.6.4 The DFT by Directed Method

In comparison, the direct calculation would be as follows:

$$x(k) = \sum_{n=0}^{N-1} x(n) \times (W_N)^{nk} \qquad k = 0, 1, \ldots N-1 \qquad W_N = e^{-\frac{j2\pi}{N}}$$

$$X(0) = x(0)W^0 + x(1)W^0 + x(2)W^0 + x(3)W^0 + x(4)W^0 + x(5)W^0 + x(6)W^0 + x(7)W^0$$
$$= (1 + j \times 0) + (0.707 + j \times 0) + (0 + j \times 0) + (-0.707 + j \times 0) + (-1 + j \times 0)$$
$$+ (-0.707 + j \times 0) + (0 + j \times 0) + (0.707 + j \times 0) = 0 + j \times 0$$

$$X(1) = x(0)W^0 + x(1)W^1 + x(2)W^2 + x(3)W^3 + x(4)W^4 + x(5)W^5 + x(6)W^6$$
$$+ x(7)W^7 = (1 + j \times 0) + (.5 + j \times 0) + (0 + j \times 0) + (0.5 + j \times 0)$$
$$+ (1 + j \times 0) + (0.5 + j \times 0) + (0 + j \times 0) + (0.5 + j \times 0) = 4 + j \times 0$$

$$X(2) = x(0)W^0 + x(1)W^2 + x(2)W^4 + x(3)W^6 + x(4)W^8 + x(5)W^{10} + x(6)W^{12}$$
$$+ x(7)W^{14} = (1 + j \times 0) + (0 + j \times 0) + (0 + j \times 0) + (0 + j \times 0)$$
$$+ (-1 + j \times 0) + (0 + j \times 0) + (0 + j \times 0) + (0 + j \times 0) = 0 + j \times 0$$

$$X(3) = x(0)W^0 + x(1)W^3 + x(2)W^6 + x(3)W^9 + x(4)W^{12} + x(5)W^{15} + x(6)W^{18}$$
$$+ x(7)W^{21} = (1 + j \times 0) + (-0.5 + j \times 0) + (0 + j \times 0) + (-0.5 + j \times 0)$$
$$+ (1 + j \times 0) + (-0.5 + j \times 0) + (0 + j \times 0) + (-0.5 + j \times 0) = 0 + j \times 0$$

$$X(4) = x(0)W^0 + x(1)W^4 + x(2)W^8 + x(3)W^{12} + x(4)W^{16} + x(5)W^{20} + x(6)W^{24}$$
$$+ x(7)W^{28} = (1 + j \times 0) + (-0.707 + j \times 0) + (0 + j \times 0) + (0.707 + j \times 0)$$
$$+ (-1 + j \times 0) + (0.707 + j \times 0) + (0 + j \times 0) + (-0.707 + j \times 0) = 0$$
$$+ j \times 0$$

$$X(5) = x(0)W^0 + x(1)W^5 + x(2)W^{10} + x(3)W^{15} + x(4)W^{20} + x(5)W^{25} + x(6)W^{30}$$
$$+ x(7)W^{35} = (1 + j \times 0) + (-0.5 + j \times 0) + (0 + j \times 0) + (-0.5 + j \times 0)$$
$$+ (1 + j \times 0) + (-0.5 + j \times 0) + (0 + j \times 0) + (-0.5 + j \times 0) = 0 + j \times 0$$

$$X(6) = x(0)W^0 + x(1)W^6 + x(2)W^{12} + x(3)W^{18} + x(4)W^{24} + x(5)W^{30} + x(6)W^{36}$$
$$+ x(7)W^{42} = (1 + j \times 0) + (0 + j \times 0) + (0 + j \times 0) + (0 + j \times 0)$$
$$+ (-1 + j \times 0) + (0 + j \times 0) + (0 + j \times 0) + (0 + j \times 0) = 0 + j \times 0$$

$$X(7) = x(0)W^0 + x(1)W^7 + x(2)W^{14} + x(3)W^{21} + x(4)W^{28} + x(5)W^{35} + x(6)W^{42}$$
$$+ x(7)W^{49} = (1 + j \times 0) + (0.5 + j \times 0) + (0 + j \times 0) + (0.5 + j \times 0)$$
$$+ (1 + j \times 0) + (0.5 + j \times 0) + (0 + j \times 0) + (0.5 + j \times 0) = 0 + j \times 0$$

13.6.5 The DFT by FFT Method

$$X(k) = \sum_{n=0}^{N-1} x(n) W_N^{nk} = \sum_{n=0}^{7} x(n) W_8^{nk}; \qquad W_8 = e^{\frac{-j2\pi}{8}}; \quad k = 0, 1, \ldots 7$$

$$= \sum_{n=0}^{3} x(2n) W_4^{nk} + W_8^k \sum_{n=0}^{3} x(2n+1) W_4^{nk}; \qquad W_4 = e^{\frac{-j2\pi}{4}}; \quad k = 0, 1, \ldots 3$$

$$= \sum_{n=0}^{1} x(4n) W_2^{nk} + W_4^k \sum_{n=0}^{1} x(4n+2) W_2^{nk}$$

$$+ W_8^k \left[\sum_{n=0}^{1} x(4n+1) W_2^{nk} + W_4^k \sum_{n=0}^{1} x(4n+3) W_2^{nk} \right]; W_2 = e^{\frac{-j2\pi}{2}}; \quad k = 0, 1$$

Periodicity property of Twiddle factor used in the FFT implies:

$$W_N^K = W_N^{N+K} \quad \text{or} \quad W_8^K = W_N^{8+K}$$

Therefore, one needs to calculate the twiddle factor for $W_8^0 \to W_8^7$.

13.7   SUMMARY

In summary, the usefulness of the FFT is the same as the DFT in Power Spectral Analysis or Filter Simulation on digital computers; however, the FFT is fast and requires fewer computations than the DFT. Decimation-in-time is the operation of separating the input data series, $x(n)$ into two $N/2$ length sequences of even-numbered points and of odd-numbered points, which can be done as long as the length is an even number, i.e., 2 to any power. Results of the decimation in the FFT are better shown with the "Butterfly"

Index	Real (a_n)	Imaginary(jb_n)	
0	2	0	DC offset
1	0.5	0.95	Fundamental (f_0)
2	3	2.1	2nd Harmonic
3	1.1	−6.2	$3 \times f_0$
4	0.25	−6.7	Center fold
5	1.1	6.2	Conjugate $3 \times f_0$
6	3	−2.1	Conjugate $2 \times f_0$
7	0.5	−0.95	Conjugate f_0

FIGURE 13.6: Output values for an eight-sample FFT

signal flow diagram with the number of complex computation Stages as, $V = Log2N$, since $N = 2V$. In general, computation of a subsequent stage from the previous butterfly is given by the basic butterfly equations, which reduce the number of complex operations from N^2 for the DFT to $NLog2N$ for the FFT; where $N = 2^m$, and m is a positive integer. The output of the FFT is shown in Fig. 13.6. Note that the first index output is the DC value of the signal. Note that the spectrum folds about the center (index 4) and the output values for indices, 5 through 7, are the conjugate of the output values for indices, 1 through 3.

CHAPTER 14

Truncation of the Infinite Fourier Transform

Problems in the accuracy of the Fourier Transform occur when adherence to the "Dirichlet Conditions" have not been met. Recall the Dirichlet condition requirements on the signal to be transformed as

1. a finite number of discontinuities,

2. a finite number of maxima and minima, and

3. that the signal be absolutely convergent (14.1)

$$\int_o^T \left| f(t) \right| dt < \infty \qquad (14.1)$$

It is impractical to solve an infinite number of transform coefficients; therefore, the engineer must decide how many Fourier coefficients are needed to represent the waveform accurately. How to make a decision on where to truncate the series is a very important part of signal processing.

Truncation means that all terms after the nth term are dropped, resulting in n finite number of terms to represent the signal. However, not including all the coefficients results in an error referred to as the "Truncation error." The truncation error, ε_n, is defined as the difference between the original function $f(t)$ and the partial sum, $s_n(t)$ of the inverse transformed truncated Fourier terms.

$$\varepsilon_n = f(t) - s_n(t)$$

where ε_n is the truncation error. The partial sum is often expressed as the "mean-squared error" as in (14.2).

$$E_n = \frac{1}{T} \int_0^T [\varepsilon_n(t)]^2 dt \qquad (14.2)$$

To understand truncation of a series, it is important to know what is desirable in truncation. Since accuracy is one of the most desirable specifications in any application, it would be desirable to attain acceptable accuracy with the least number of terms. Without proof in limits and convergence, let us define convergence as the rate at which the truncated series approaches the value of the original function, $f(t)$. In general, the more rapid the convergence, the fewer terms are required to obtain desired accuracy.

The convergence rate of the Fourier Transform is directly related to the rate of the decrease in the magnitude of the Fourier Coefficients. Recall the Fourier transform coefficients of a square wave as having the following magnitudes:

$$|a_3| = \frac{1}{3}; \quad |a_5| = \frac{1}{5}; \quad |a_7| = \frac{1}{7}; \quad |a_9| = \frac{1}{9}; \cdots; \quad |a_n| = \frac{1}{n}$$

Note that the sequence is the reciprocal of the nth term to the 1st power.

Now let us examine the Fourier transform of the triangular waveform shown in Fig. 14.1.

The Fourier trigonometric series representation is given by (14.3).

$$v(t) = \frac{8V}{\pi^2} \left(\sin \omega_0 t - \frac{1}{3^2} \sin 3\omega_0 t + \frac{1}{5^2} \sin 5\omega_0 t + \cdots + \frac{1}{n^2} \sin n\omega_0 t \right) \qquad (14.3)$$

Note that the magnitude of the coefficients are the square of the nth term.

$$|a_3| = \frac{1}{3^2}; \quad |a_5| = \frac{1}{5^2}; \quad |a_7| = \frac{1}{7^2}; \quad |a_9| = \frac{1}{9^2}; \cdots; \quad |a_n| = \frac{1}{n^2}$$

It is noted that the Fourier coefficients for a triangular waveform diminish faster (rate of convergence) than the Fourier coefficients for a rectangular waveform. To help in

FIGURE 14.1: One cycle of a triangular waveform with odd symmetry

TABLE 14.1: Convergence Table

$f(t)$	JUMP-IN	IMPULSE-IN	DECREASING a_n	WAVEFORM
Square wave	$f(t)$	$f'(t)$	$1/n$	Discontinuities
Triangle wave	$f'(t)$	$f''(t)$	$1/n^2$	
Parabolic/ sinusoid	$f''(t)$	$f'''(t)$	$1/n^3$	
...				
---	$f^{k-1}(t)$	$f^k(t)$	$1/n^k$	Smoother

determining the rate of convergence, a basic expression was formulated called the Law of Convergence.

The law covering the manner in which the Fourier coefficients diminish with increasing "n" is expressed in the number of times a function must be differentiated to produce a jump discontinuity. For the kth derivative, the convergence of the coefficients will be on the order of $1/n^{k+1}$. For example, if that derivative is the kth, then the coefficients will converge at the rate shown in (13.4).

$$|a_n| \leq \frac{M}{n^{k+1}} \quad \text{and} \quad |b_n| \leq \frac{M}{n^{k+1}} \tag{13.4}$$

where M is a constant, which is dependent on $f(t)$.

There are two ways to show convergence. Let us look at Table 14.1, the Convergence Table. The left most column of the table denotes the type of waveform with the right most column describing the waveform as going from with discontinuities to smoother. The second and third columns labeled "Jump-in" and "Impulse-in," respectively. *Jump-in* means that discontinuity occurs in the function after successive differentiation, where the discontinuities occur in the *square-like waveform*. *Impulse-in* means what derivative (successive differentiation) of the function will result in an *impulse train*. The fourth column gives the general expression for convergence of the coefficients. From the table one may conclude that the smoother the function, the faster the convergence.

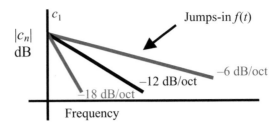

FIGURE 14.2: Graphs of convergence rates

Often, the rates of decrease in the Fourier coefficients with the number of terms, n, are expressed in the logarithmic terms of decibels per octave by expressing the absolute magnitude of the Fourier coefficients, $|c_n|$ in decibels and plotting the coefficients against frequency as shown in Fig. 14.2.

14.1   PRACTICAL APPLICATIONS

So how realistic are discontinuities in the real world, specifically physiological signals in clinical and biomedical engineering? Unfortunately, discontinuities in medicine are real. Discontinuities in the human and animal physiological waveforms may be found in the following signals:

1. Electroencephalograms (EEG); Seizure waveforms

2. Spike waveforms in EEG and Electromyogram (EMG)

3. Saccadic movements in the Electro-occulograms (EOG)

It is important to understand what problems discontinuities present in signal processing. It is well known that as more coefficients are added to the waveform approximation, the accuracy improves everywhere except in the immediate vicinity of a discontinuity. Even in the "Limit" as $n \to \infty$, the discrepancy (error) at the point of discontinuity becomes approximately 9%. This error is known as the *Gibbs phenomenon*.

One may think of the Gibbs phenomenon as the response to a square pulse, which is the characteristic overshoot and damped oscillatory decay in nonlinear systems (filters). What the discontinuities and the Gibbs phenomenon indicate is that one can expect about a 9% error at each region of discontinuity in a signal. It should be noted that discontinuities may occur at the beginning and end points of the analysis window.

Can one correct for the discontinuities and the Gibbs phenomenon? There are a few approaches that may be used. For example, one could use the low-pass filter the signal to remove the high-frequency components, thus smoothing the signal, for example, rounding off the QRS in the electrocardiogram (ECG) signal. Or, one could avoid any analysis in the region of discontinuity by "Piece-meal" or segment analysis.

14.2 SUMMARY

Keep in mind that all Fourier analysis are approximations due to truncation in the number of Fourier coefficients; therefore, from "lessons learned," one should do the best approximation possible, knowing that 100% accuracy is not attainable. The smoother the signal, the faster it will converge, thus requiring fewer Fourier coefficient terms. Discontinuities either at the edges or in the middle of the signal will result in greater discrepancy (error). Remember that the Gibbs phenomenon indicates that approximately a 9% error will occur in the region of any discontinuity.

C H A P T E R   1 5

Spectral Analysis

15.1 INTRODUCTION

Spectral analysis is the process by which we estimate the energy content of a time-varying function (or signal) as a function of frequency. For signals with very little frequency content, the spectral density function can be represented by a "line spectrum." The variations in the frequency resolution of spectra are due to different window lengths. As the signal becomes more complex (in terms of frequency content), the power spectrum resembles a more continuous function, hence, spectral density functions. The sharp peaks that correspond to the raw estimates of the energy for the main frequency components will be present; however, a certain amount of signal conditioning (i.e., smoothing) is necessary prior to this final estimation of the "Smoothed Power Spectral Estimate." Notice that at times, the text will refer to the Spectral Density Function as the "Power Spectrum" (singular) or "Power Spectra" (plural). Units for the power spectra are in power per frequency band (spectral resolution, $\Delta f = 1/T$).

The harmonic analysis of a random process will yield some frequency information, whereas the power spectrum is a useful tool in analysis of a random process. So let us examine what information can be obtained from the power spectrum. The power spectrum has four major uses. The first is to get an overview of the function's frequency distribution. Second is for comparison of distribution via statistical testing or by discriminate analysis. Third, the power spectrum can be used to obtain estimates the parameters of interest. And last, but probably the most important application is that power spectra can be used to show hidden periodicities. In summary, spectral estimates are used primarily to

1. show hidden *Periodicities* in frequency domain;

2. obtain descriptive statistics, for example, distribution, mean, mode, bandwidth of the spectral from a random signal;

3. get an overview of the frequencies in a function, $F(\omega)$;

4. in obtaining parameter estimation and/or feature extraction, that is, clinical EEG bands; and

5. in classification testing and discrimination analysis.

There are several methods for calculating the power spectra. The most commonly used spectral estimation methods are calculated via

1. the autocorrelation function,

2. the direct or fast fourier transform,

3. the autoregression, which is briefly included in this chapter, but this method of spectral estimation will not be covered in the course, since an autoregression course is offered by the statistics department.

It should be noted, however, that there are other methods. The power spectrum (periodogram), most commonly used by engineers, is an estimator of the "raw power spectrum" and must undergo smoothing to clearly reveal informative features (preferred by statisticians). Autoregressive spectral estimation is a powerful tool that estimates the power spectrum without employing any smoothing techniques.

Before commencing with the spectral analysis of a signal, the signal has to be qualified as deterministic (periodic) or random. The random data must be tested for normality of distribution, independence, and stationarity. If any of these statistical properties are not met, the data is a nonstationary process. If the data is nonstationary, appropriate steps should be followed to make the data independent and stationary before proceeding with spectral analysis; however, nonstationary processes are beyond the scope of this book.

15.2 SPECTRAL DENSITY ESTIMATION

A basic question one might ask is whether the Fourier Series or transform is applied to a voltage signal, how did the unit become power? To answer this question, let us consider the function $v(t)$ defined over the interval $(-\infty, +\infty)$. If $v(t)$ is periodic or almost periodic, it can be represented by the Fourier series as in (15.1).

$$v(t) = a_0 + \sum_{1}^{\infty} (a_n \cos nwt + b_n \sin nwt) \tag{15.1}$$

where $\omega = 2\pi f$, f represents the frequency, and the period is $T = 1/f$. The complex Fourier coefficients a and b are then represented by equations in (15.2).

$$a_0 = \frac{1}{2T} \int_0^T v(t)\,dt$$

$$a_n = \frac{1}{2T} \int_0^T v(t)\,\cos(nw_0 t)\,dt \tag{15.2}$$

$$b_n = \frac{1}{2T} \int_0^T v(t)\,\sin(nw_0 t)\,dt$$

The magnitude of the complex coefficients are obtained from (15.3), which has units of "Voltage/Hertz."

$$c_n = \sqrt{(a_n + b_n)(a_n - b_n)} \tag{15.3}$$

The Fourier coefficients are not in units of energy or power. By using Parseval's Theorem, the power spectrum is obtained from (15.4).

$$c_n^2 = a_n^2 + b_n^2 \tag{15.4}$$

The squared magnitude of the complex coefficients from (15.4) has units of "Voltage Squared/Hertz," "Average energy or power per hertz."

Parseval's Theorem may be stated as the following equation form (15.5).

$$\int_{-\infty}^{+\infty} |v(t)|^2\,dt = \frac{1}{2\pi} \int_{-\infty}^{+\infty} |X(\omega)|^2\,d\omega \tag{15.5}$$

where $v(t)$ and $X(\omega)$ are the Fourier Transform pair.

The left-hand side of (15.5), is defined as the total energy in the signal $v(t)$.

A periodic signal contains real and imaginary components of power. However, determination of even symmetry (real power only) and odd symmetry (imaginary power only) simplifies calculations because b_n and a_n cancel for each case, respectively. Fourier Series Analysis is a very powerful tool, but it is not applicable to random signals.

15.2.1 Power Spectra Estimation Via the Autocorrelation Function

The classic method used to efficiently obtain the estimated Power Spectra from random processes before the advent of the Fast Fourier Transform was by taking the finite Fourier transform of the correlation function. The process required one to calculate the autocorrelation function (or cross-correlation) first, and then to take the Fourier Transform of the resulting correlation function.

The autocorrelation function (R_{xx}) and cross-correlation function (R_{xy}) will not be discussed in this section since the material was presented in Chapter 9. Hence, let us assume that the correlation function exists and that its absolute value is finite. The Fourier Transform (FT) is defined as shown in (15.6) for the autocorrelation and (15.7) for the cross-correlation:

$$S_{xx}(\omega) = \int_{-\infty}^{\infty} R_{xx}(\tau)e^{-j\omega\tau}\,d\tau \qquad (15.6)$$

$$S_{xy}(\omega) = \int_{-\infty}^{\infty} R_{xy}(\tau)e^{-j\omega\tau}\,d\tau \qquad (15.7)$$

where $S_{xx}(\omega)$ is the autospectral density function for an input function $x(t)$, and $S_{xy}(\omega)$ is the cross-spectral density function for the input $x(t)$ and output $y(t)$.

The cross-spectral density function is estimated by taking the cross-correlation of two different functions in the time domain, that is, the input signal to a system and the output signal from the same system, and then taking the Fourier transform of the resulting cross-correlation function, S_{xy}. The Inverse Fourier Transforms are then

$$R_{xx}(\tau) = \frac{1}{2\pi} \int_{-\infty}^{\infty} S_{xx}(\omega)e^{j\omega t}\,d\omega \qquad (15.8)$$

$$R_{xy}(\tau) = \frac{1}{2\pi} \int_{-\infty}^{\infty} S_{xy}(\omega)e^{j\omega t}d\omega \tag{15.9}$$

Equations (15.6) through (15.9) are called the "Wiener–Kinchine Relation."

The Fourier transform of the correlation functions contain real and imaginary components of energy, meaning that the spectrum is a two-sided spectrum with half of the energy containing negative frequencies, that is, $-\omega$; hence, the following equivalent equations are defined in (15.10) and (15.11).

$$S_{xx}(-\omega) = \overline{S}_{xx}(\omega) = S_{xx}(\omega) \tag{15.10}$$

$$S_{xy}(-\omega) = \overline{S}_{xy}(\omega) = S_{xy}(\omega) \tag{15.11}$$

where the bar over the letter S denotes the complex conjugate.

The $S(-\omega)$ and $S(\omega)$ are the estimate of the "two-sided spectral density function," in which the magnitude of the FT for a frequency ω in the real plane is equal to the magnitude of the FT of ω in the imaginary plane. In the case when the autocorrelation function is evaluated over the interval from 0 to ∞, the autospectra may be obtained by using the symmetry properties of the function. An alternate equation is given in (15.12), with its inverse transform as (15.13).

$$S_{xx}(\omega) = 2 \int_{0}^{\infty} R_{xx}(\tau) \cos(\omega\tau)d\tau \tag{15.12}$$

$$R_{xx}(\tau) = \frac{1}{\pi} \int_{0}^{\infty} S_{xx}(\omega) \cos(\omega\tau)d\omega \tag{15.13}$$

For the case of the "one-sided spectral density function" (Fig. 15.1), in which only the real frequency components of the power spectrum are represented, (15.14) and (15.15) are the same as doubling the energy from one side of the two-sided spectra.

$$G_{xx}(\omega) = 2S_{xx}(\omega) \tag{15.14}$$

$$G_{xy}(\omega) = 2S_{xy}(\omega) \tag{15.15}$$

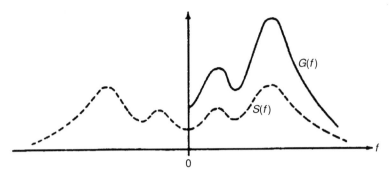

FIGURE 15.1: The two-sided and one-sided spectral density functions

15.2.2 Power Spectra Estimation Directly from the Fourier Transform

With computers and the Fourier transform, it is easier to obtain the spectral estimate directly from the either discrete Fourier transform (DFT) or the fast Fourier transform (FFT). The FFT has become the most popular and widely used method for spectral estimation. The first step in the process requires one to obtain the Fourier coefficients, and then to multiply each coefficient of the autospectrum by its conjugate to get spectral power. The cross-spectral density function is obtained by taking the Fourier transform of the two functions, $x(t)$ and $y(t)$, then the coefficients of one function, $X(\omega)$ are multiplied by the conjugate of the coefficients of the second function, $Y(\omega)$. The equations for the autospectum ($\hat{G}_{xx}$) and the cross-spectrum ($\hat{G}_{xy}$) are given by (15.16) and (15.17), respectively. The "^" symbol is used to denote a "raw" spectrum.

$$\hat{G}_{xx} = F_1(\omega) \times \overline{F_1}(\omega) \tag{15.16}$$

where the "bar" denotes the conjugate of the function.

$$\hat{G}_{xy} = F_1(\omega) \times \overline{F_2}(\omega) \tag{15.17}$$

An alternate method that works well, but only with the autospectrum, is to obtain the Fourier coefficients and calculate the sum of the real term squared, a_n^2, and the imaginary term squared, b_n^2, as in (15.18). Equation (15.18) should not be used to calculate the cross-spectrum.

$$\hat{G}_{xx} = a_n^2 + b_n^2 \tag{15.18}$$

15.3 CROSS-SPECTRAL ANALYSIS

Cross-spectral analysis is the process for estimating the "*Energy*" content of "Two" time-varying functions (signals) as a function of frequency. The cross-spectrum has many practical uses in engineering among which the major applications are in

1. determining output/input relationship,

2. obtaining the transfer function of a system from its input and output spectra,

3. determining bode plots from the spectra
   a. Magnitude, $H(\omega)$

   b. Phase angle, $\theta(\omega)$,

4. estimating transfer function parameter, and

5. determining "Coherence" or *Goodness Fit* of the Transfer Function (Model).

The equation for the "one-sided cross-spectral function" from the cross-correlation method is given by (15.19).

$$G_{XY}(f) = 2S_{XY}(f) \quad 0 \le f < \infty \tag{15.19}$$

The expanded form of (15.19) may be expressed as (15.20).

$$G_{XY}(f) = \int_{-\infty}^{\infty} R_{XY}(\tau)e^{-j2\pi f\tau}d\tau = C_{XY}(f) + jQ_{XY}(f) \tag{15.20}$$

Complex terms of the cross-spectrum are

1. $C_{xy}(f)$ is called the *coincident spectral density function (Co-spectrum),* and

2. $Q_{xy}(f)$ is called the *quadrature spectral density function (Quad-spectrum).*

Since the cross-spectrum has two terms, a real and an imaginary term, the cross-spectrum can be represented in complex polar notation as in (15.21).

$$G_{XY}(f) = |G_{XY}(f)| e^{-j\theta_{xy}(f)} \quad \text{for} \quad 0 \le f < \infty \tag{15.21}$$

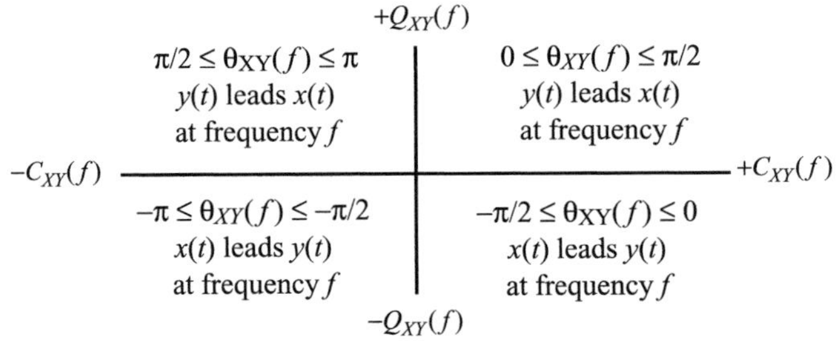

$$+Q_{XY}(f)$$

$\pi/2 \leq \theta_{XY}(f) \leq \pi$	$0 \leq \theta_{XY}(f) \leq \pi/2$
$y(t)$ leads $x(t)$ at frequency f	$y(t)$ leads $x(t)$ at frequency f

$-C_{XY}(f)$ ———————————————————————— $+C_{XY}(f)$

$-\pi \leq \theta_{XY}(f) \leq -\pi/2$	$-\pi/2 \leq \theta_{XY}(f) \leq 0$
$x(t)$ leads $y(t)$ at frequency f	$x(t)$ leads $y(t)$ at frequency f

$$-Q_{XY}(f)$$

FIGURE 15.2: The relationship between the phase angle to cross-spectral terms

The magnitude of the cross-spectrum is calculated from (15.22), and the phase is calculated from (15.23).

$$\left| G_{XY}(f) \right| = \sqrt{C_{XY}^2(f) + Q_{XY}^2} \qquad (15.22)$$

$$\theta_{XY}(f) = \tan^{-1} \frac{Q_{XY}(f)}{C_{XY}(f)} \qquad (15.23)$$

Special note should be made of the fact that the autospectrum does not retain any phase information whereas the cross-spectrum contains phase information. Keep in mind that engineers look for relationships in both magnitudes or in phase between an input stimulus and the output response of a system. Phase relations are often expressed as leading or lagging a reference signal. The relationship between the phase angle to cross-spectral terms is shown in Fig. 15.2. Since phase is calculated with a tangential function, the relationship is expressed in quadrants.

15.4   PROPERTIES OF SPECTRAL DENSITY FUNCTIONS

For both autospectra and cross-spectra, the Power Spectral Density functions are "*Even*" Functions and the Power Spectra are always "Positive" valued functions, $G_{xx} > 0$.

15.5   FACTORS AFFECTING THE SPECTRAL DENSITY FUNCTION ESTIMATION

There are several factors that adversely affect the spectral density function estimation, regardless of the method of estimation that is employed, if these factors are not taken into account.

1. *Filtering*: Analog filtering is the first major consideration for the acquisition and processing of data. The engineer must determine the appropriate cutoff frequencies within the signal. Both high- and low-frequency cutoff should be determined and appropriate analog filters with sufficient attenuation in the bandstop regions should be employed. The high-pass filter should be used to remove the DC component of the signal and to remove trends. An antialias filter (low-pass filter) should be employed to eliminate "Aliasing" of the data. It is important to use an analog antialias filter prior to sampling of the data; any subsequent digital filtering should be done after sampling.

2. *Frequency resolution*: Frequency resolution is defined as the smallest unit of frequency in the spectrum. The "Fundamental Frequency" or resolution of the spectrum $(\Delta f_0 = 1/T)$ is equal to the inverse of time the signal segment (length of time in seconds: T) that is being analyzed by Spectral Analysis. The segment time is also referred to as the window of observation or analysis. The energy in the spectrum of the signal is identified as harmonics of the fundamental frequency, that is, multiples of the fundamental frequency.

3. *Digitizing and the Nyquist criterion*: Probably the most important rule to follow in spectral analysis is the Nyquist Criteria in digital sampling of the signal. The sampling rate must be twice the highest frequency of the signal bandwidth (W). Sampling at a frequency lower than the Nyquist frequency causes aliasing of the spectrum, which will then yield an erroneous spectral estimator.

4. *Windowing*: A "Window" function should be applied prior to obtaining the estimation of the "Spectral Density Function." Windowing is a process by which the sampled data points are "weighted" in the time domain. The effect is smoothing of the signal at the beginning and end of the record to prevent the occurrence of discontinuities and Gibb's Phenomenon in the data. Improper windowing produces "Leakage" in the spectrum, which adversely affects estimation of the Spectral Density Function. There are numerous window functions that can be applied. The problem lies in deciding which window function is the best to apply on the signal being studied.

5. *Smoothing*: To obtain a fairly good estimator, statisticians would insist on some "Smoothing" function, since the Spectral Density Function or Periodogram is a raw power spectrum estimate that contains high-frequency components. Smoothing is a technique equivalent to low-pass filtering to eliminate the high-frequency components from the spectrum for the final smoothed spectral density estimate.

15.6 ADVANCED TOPIC

15.6.1 Brief Description of Autoregressive Spectral Estimation

The concept of smoothing serves as a transition to the Autoregressive (AR) Spectral Estimation Method. The AR Spectral Estimator, also known as the Maximum Entropy Spectral Estimator, is becoming more popular in the engineering community as an accurate estimator of the spectral density function.

To perform the AR Spectral Estimation, it is necessary to do the following procedural steps in order.

1. Fit AR model to data

2. Calculate AR coefficients

3. Enter the AR coefficients into formula for spectrum of AR process.

4. Estimate spectrum of AR process

The AR model of order p is represented by (15.24):

$$x_t = \alpha_0 + \alpha_1 x_{t-1} + \alpha_2 x_{t-2} + \cdots + \alpha_p x_{t-p} + z_t \tag{15.24}$$

The order of the model is chosen using Akaike's Information Criterion (AIC). The coefficients are then calculated from (15.25).

$$\hat{\alpha}_k = \frac{\sum_{t=1}^{N-1} (x_t - \bar{x})(x_{t-1} - \bar{x})}{\sum_{t=1}^{N-1} (x_t - \bar{x})^2} \tag{15.25}$$

where $\bar{x}$ is the mean of the sample record, N is the total number of data points, $x(t)$ is the point at time t, and $\hat{\alpha}_k$ is the approximate estimator of α.

Another approximation that has often been used is given by (15.26).

$$\sum_{t=1}^{N-1} (x_t - \bar{x})^2 = \sum_{t=1}^{N} (x_t - \bar{x})^2 \qquad (15.26)$$

If (15.26) is used then $\alpha_k \cong r_k$, where r_k is the kth correlation coefficient. Then the sample correlation coefficients are substituted into the Yule–Walker equations (15.27), and the matrix equations are solved for α, the AR coefficients.

$$
\begin{bmatrix}
1 & r_1 & & & r_{p-1} \\
r_2 & 1 & r_1 & & r_{p-2} \\
- & - & - & - & - \\
- & - & - & - & - \\
- & - & - & - & r_1 \\
r_{p-1} & - & - & - & 1
\end{bmatrix}
x
\begin{bmatrix}
\alpha_1 \\
\alpha_2 \\
- \\
- \\
- \\
\alpha_p
\end{bmatrix}
=
\begin{bmatrix}
r_1 \\
r_2 \\
- \\
- \\
- \\
r_p
\end{bmatrix}
\qquad (15.27)
$$

Since the Fourier transform is given by (15.28), the autospectral density function is estimated by using an AR model of order p as given in (15.29).

$$F(\omega) = \sum_{k=-\infty}^{+\infty} \gamma_k e^{-j\omega k} \qquad (15.28)$$

$$X(\omega) = \frac{\sigma^2}{\pi}\left(1 + \sum_{k=1}^{p} \alpha_k e^{-j\omega k} + \sum_{k=1}^{p} \alpha_k e^{j\omega k}\right) \qquad (15.29)$$

where the autocovariance function is $\gamma(k) = \sigma^2 \alpha_k$.

It should be noted that the frequency ω varies within the interval $(0, \pi)$, where π represents the Nyquist Frequency.

15.6.2 Summary

The cross-spectrum shows the relationship between two waveforms (input and output) in the "Frequency Domain." The cross-spectrum indicates what frequencies in the output are related to frequencies in the input. The cross-spectrum consists of complex terms, which are termed the *coincident spectral density function* [*Co-spectrum, Cxy(f)*] and the *quadrature spectral density function [Quad-spectrum Qxy(f)].* The cross-spectrum is calculated in the same manner as the autospectrum, via direct Fourier Transform of the two signals or via the cross-correlation function. As with the autospectrum, the "Raw" cross-spectrum must be *smoothed*, and then *normalized* to get a good estimation of the cross-spectrum.

15.7   SUGGESTED READING

1. Bendat, J., and Piersol, A. G. *Random Data Analysis and Measurement Procedures.* New York: Wiley & Sons, Inc., 1986.

2. Chatfield, C. *The Analysis of Time Series: An Introduction.* London: Chapman & Hall, 1980.

3. Lynn, P. A. *An Introduction to the Analysis and Processing of Signals.* New York: Wiley & Sons, Inc., 1973.

4. McIntire, D. A. "A Comparative Case Study of Several Spectral Estimators," In *Applied Time Series Analysis*, D. F. Findley, ed., Academic Press, 1978.

CHAPTER 16

Window Functions and Spectral Leakage

The purpose of window functions is to minimize a phenomenon called "Spectral Leakage," which is the result of selecting a finite time interval and a nonorthogonal trigonometric basis function over the interval of analysis. Only frequencies that coincide with the basis function will project on a single basis vector, such as shown in Fig. 16.1. Signals, which include frequencies not on the basis set, will not be periodic in the observing interval (often referred to as the "Observation or Observing Window"). The periodic extension of a signal with a period that is not equal to the fundamental period of the window (as in Fig. 16.2) will produce discontinuities at the boundaries of the observing interval. It is these discontinuities that cause the frequency leakage of the signal across the spectrum.

The sine function in Fig. 16.2 does not fit perfectly in the window, since its value at the beginning is not the same as the value at the end of the observation interval [4]. Since the window is supposed to contain a periodic function, one should note that in the next cycle, the value must go from 1 to 0 at the same instant of time. Hence, the discontinuity at the edge of the window will result in the spectral power of a specific frequency leaking out of the main lobe or specific spectral bandwidth to adjacent spectral side lobes. Leakage into the side lobes may occur in either side (direction) of the main lobe. The decrease of power from the main lobe adds to the power in the sideband harmonics resulting in an erroneous estimate of power in the sideband harmonics.

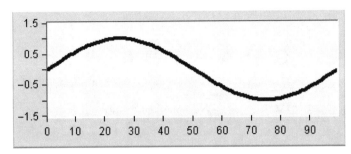

FIGURE 16.1: Sinusoidal waveform. The sine function fits perfectly in the window, since its value is zero (0) at the beginning and end of the observation interval

Outward leakage can be easily demonstrated by some simple examples. Figure 16.3 shows the Power Spectral Density (PSD) of a sinusoidal with a period that is equal to the observation window length of 1 second. Note that there is only one lobe at 1 Hz.

Figure 16.4 shows the Power Spectral Density (PSD) of a sinusoidal with a period that is not equal to the observation window length of 1 second. Note the main-lobe spectral energy at 0.59 Hz with side lobes at 0.41, 0.50, 0.68, and 0.77 Hz.

In dealing with random physiological data, the possibility of an integral fundamental frequency repetition is very limited; hence, regardless of what sampling window interval is used, leakage will be present in the spectra.

16.1   GENERALITIES ABOUT WINDOWS

The window functions (often called restoring functions) are basically weighting functions applied to the time domain data. Weighting function selection can be made early in the design process because the choices of FFT algorithm and weighting function are independent of each other. Choice of a weighting function to provide the specified

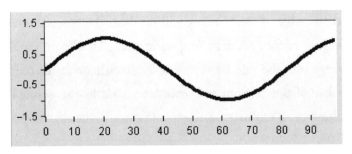

FIGURE 16.2: Discontinuity in sinusoidal waveform

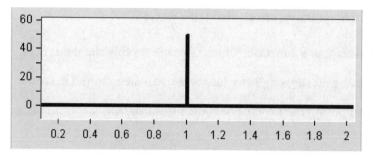

FIGURE 16.3: Power spectral density of the trace in Fig. 16.1. Note the main lobe at 1 Hz without any side lobes

side-lobe level is done without concern for the FFT algorithm that will be used because

1. they work for any length FFT,

2. they work the same for any FFT algorithm, and

3. they do not alter the FFT ability to distinguish two frequencies (resolution).

Window functions, $w(t)$, have the following properties:

1. The $w(t)$ function has to be real, even and nonnegative.

2. The transformed function has to have a main lobe at the origin of the fundamental frequency, and side lobes at both sides to ensure symmetry.

3. If the nth derivative has an "impulse" character, then the peak of the side lobes of the transformed function will decay asymptotically as $6*n$ dB/octave (least estimate).

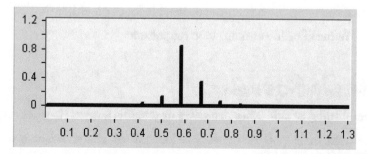

FIGURE 16.4: Power spectral density of the trace with discontinuity shown in Fig. 16.2. Note the main lobe of the spectra at 0.59 Hz with side lobes at 0.41, 0.50, 0.68, and 0.77 Hz

Weighting functions are applied three ways:

1. as a rectangular function, which does not modify the input data,

2. by having all the weighting function coefficients stored in memory, and

3. by computing each coefficient when it is needed.

16.2   PERFORMANCE MEASURES

The choice of a weighting function depends on which of the features of the narrowband DFT filters (*resolution*) are most important to the application. Those features are performance measures of the narrowband filters to analytically compare weighting functions. All these measures, except frequency straddle loss, refer to individual filters. Frequency straddle loss is associated with how filters work together [3].

16.2.1 Highest Side-Lobe Level

Highest side-lobe level (in decibels) is an indication on how large the effect of the side lobes is. Side lobes are a way of describing how a filter responds to signals at frequencies that are not in its main lobe, commonly called its passband. Each FFT filter has several side lobes. With rare exception, the highest side lobe is closest in frequency to the main lobe and is the one that is most likely to cause the passband filter to respond when it should not. For a signal with the maximum point of the main lobe at 0 dB, rectangular window the first side lobe is about −13 dB, relative amplitude. A good leakage suppressor should have its first side lobe much lower than −13 dB. Since side lobes are the result of leakage from the main lobe, then the more side lobes on either side of the main lobe will result in a reduced or lower main lobe magnitude.

16.2.2 Sidelobe Fall-Off Ratio

The peak (amplitude) of side lobes decreases or remains level as they get further away in frequency from the main lobe passband. Side lobe fall-off characteristic is very important since it indicates the rate at which the side lobes are decreasing in magnitude. The rectangular window has a decreasing rate of -6 dB/octave. This means that the next side lobe will be encountered at $-13 - 6 = -19$ dB. If the side lobes reduce rapidly, then the main lobe of the FFT bands will lose less energy into the sidelobes and the spectra will be less erroneous. Side lobe fall-off performance measure is important for applications with multiple signals that are close in frequency.

16.2.3 Frequency Straddle Loss

Frequency straddle loss is the reduced output of a DFT filter caused by the input signal not being at the filter's center frequency. Frequencies seldom fall at the center of any filter's passband. When a frequency is halfway between two filters, the response of the FFT has its lowest amplitude. For a rectangular weighting function, the frequency response halfway between two filters is 4 dB lower than if the frequency were in the center of a filter. Each of the other weighting functions in this chapter has less frequency straddle loss than the rectangular one. This performance measure is important in applications where maximum filter response is needed to detect the frequencies of interest.

16.2.4 Coherent Integration Gain

Coherent integration gain (often referred to as *Coherent Gain* or *Processing Loss*) is the ratio of amplitude of the DFT filter output to the amplitude of the input frequency. N-point FFTs have a coherent gain of N for frequencies at the centers of the filter passbands. Since most weighting function coefficients are less than 1, the coherent gain of a weighted FFT is less than N. While weighting functions reduce the coherent integration gain, the combination of this reduction and the improved straddle loss results in an overall signal response improvement halfway between two filters. Like frequency straddle loss, this performance measure is important in applications where maximum filter response is needed to detect the frequencies of interest. To restore the main lobes to their original magnitudes, the FFT coefficients may be multiplied by the inverse of the coherent gain. Window coherent gains are compared as shown in Table 16.1 to the rectangular window gain, which is 1 or 0 dB.

TABLE 16.1: Windows and Figures of Merit

WINDOW	HIGHEST SIDE-LOBE LEVEL (DB)	SIDE-LOBE FALL-OFF (DB/OCT)	COHERENT GAIN	EQUIV. NOISE BW (BINS)	3.0-DB BW (BINS)	SCALLOP LOSS (DB)	WORST CASE PROCESS LOSS (DB)	6.0-DB BW (BINS)	OVERLAP CORRELATION (PCNT) 75% OL	50% OL
Rectangle	−13	−6	1.00	1.00	0.89	3.92	3.92	1.21	76.0	60.0
Triangle	−27	−12	0.50	1.33	1.28	1.82	3.07	1.78	71.9	25.0
Cos*(X) Hanning										
α = 1.0	−23	−12	0.64	1.23	1.20	2.10	3.01	1.65	75.5	31.8
α = 2.0	−32	−16	0.50	1.50	1.44	1.42	3.16	2.00	65.9	16.7
α = 3.0	−39	−24	0.42	1.73	1.66	1.08	3.47	2.32	56.7	8.5
α = 4.0	−47	−30	0.38	1.94	1.86	0.86	3.75	2.59	48.6	4.3
Hamming	−43	−6	0.64	1.36	1.30	1.78	3.10	1.81	70.7	23.5
Riesz	−21	−12	0.67	1.20	1.16	2.22	3.01	1.59	76.5	34.4
Riemann	−26	−12	0.59	1.30	1.26	1.89	3.03	1.74	73.4	27.4
De La Valle Poussin	−53	−24	0.38	1.92	1.82	0.90	3.72	2.55	48.3	5.0
Tukey										
α = 0.25	−14	−18	0.88	1.10	1.01	2.96	3.39	1.36	74.1	44.4
α = 0.50	−15	−18	0.75	1.22	1.15	2.24	3.11	1.57	72.7	36.4
α = 0.75	−19	−18	0.63	1.36	1.31	1.73	3.07	1.80	70.4	25.1
Bohman	−46	−24	0.41	1.79	1.71	1.02	3.54	2.38	54.5	7.4
Poisson										
α = 2.0	−19	−6	0.44	1.30	1.21	2.09	3.23	1.09	69.9	27.8
α = 3.0	−24	−6	0.32	1.05	1.45	1.46	3.64	2.08	64.8	15.1
α = 4.0	−31	−6	0.25	2.08	1.75	1.03	4.21	2.58	40.4	7.4

Window										
Hanning Poisson										
α = 0.5	−35	−18	0.43	1.61	1.54	1.26	3.33	2.14	61.3	12.6
α = 1.0	−39	−18	0.38	1.73	1.64	1.11	3.50	2.30	56.0	9.2
α = 2.0	NONE	−18	0.29	2.02	1.87	0.87	3.94	2.65	44.6	4.7
Cauchy										
α = 3.0	−31	−6	0.42	1.48	1.34	1.71	3.40	1.90	61.6	20.6
α = 4.0	−35	−6	0.33	1.76	1.50	1.36	3.83	2.20	48.8	13.2
α = 5.0	−30	−6	0.28	2.06	1.68	1.13	4.28	2.53	38.3	9.0
Gaussian										
α = 2.5	−42	−6	0.51	1.39	1.33	1.69	3.14	1.86	67.7	20.0
α = 3.0	−55	−6	0.43	1.64	1.55	1.25	3.40	2.18	67.5	10.6
α = 3.5	−69	−6	0.37	1.90	1.79	0.94	3.73	2.52	47.2	4.9
Dolph Cheryshev										
α = 2.5	−50	0	0.53	1.39	1.33	1.70	3.12	1.85	69.6	22.3
α = 3.0	−60	0	0.48	1.51	1.44	1.44	3.23	2.01	64.7	16.3
α = 3.5	−70	0	0.45	1.62	1.55	1.25	3.35	2.17	60.2	11.9
α = 4.0	−80	0	0.42	1.73	1.65	1.10	3.48	2.31	55.9	8.7
Kaiser Bssel										
α = 2.0	−46	−6	0.49	1.50	1.43	1.46	3.20	1.99	65.7	16.9
α = 2.5	−57	−6	0.44	1.65	1.57	1.20	3.38	2.20	59.5	11.2
α = 3.0	−69	−6	0.40	1.80	1.71	1.02	3.56	2.39	53.9	7.4
α = 3.5	−82	−6	0.37	1.93	1.83	0.89	3.74	2.57	48.8	4.8

(cont.)

TABLE 16.1: (*Continued*)

WINDOW	HIGHEST SIDE-LOBE LEVEL (DB)	SIDE-LOBE FALL-OFF (DB/OCT)	COHERENT GAIN	EQUIV. NOISE BW (BINS)	3.0-DB BW (BINS)	SCALLOP LOSS (DB)	WORST CASE PROCESS LOSS (DB)	6.0-DB BW (BINS)	OVERLAP CORRELATION (PCNT) 75% OL	50% OL
Barcilon Temes										
α = 3.0	−53	−6	0.47	1.56	1.49	1.34	3.27	2.07	63.0	14.2
α = 3.5	−58	−6	0.43	1.67	1.59	1.18	3.40	2.23	58.6	10.4
α = 4.0	−68	−6	0.41	1.77	1.68	1.05	3.52	2.36	54.4	7.6
Exact Blackman	−51	−6	0.46	1.57	1.52	1.33	3.29	2.13	62.7	14.0
Blackman	−58	−18	0.42	1.73	1.68	1.10	3.47	2.35	56.7	9.0
Minimum 3 Sample Blackman–Harris	−67	−6	0.42	1.71	1.66	1.13	3.46	1.81	57.2	9.6
Minimum 4 Sample Blackman–Harris	−92	−6	0.36	2.0	1.90	0.83	3.85	2.72	46.0	1.8
61 dB 3 Sample Blackman–Harris	−61	−6	0.45	1.61	1.56	1.27	3.34	2.19	61.0	12.6
74 dB 4 Sample Blackman–Harris	−74	−6	0.40	1.79	1.74	1.03	3.56	2.44	53.9	7.4
4 Sample Kaiser Bessel α = 3.0	−69	−6	0.40	1.80	1.74	1.02	3.56	2.44	53.9	7.4

16.2.5 Equivalent Noise Bandwidth

Equivalent noise bandwidth is the ratio of the input noise power to the noise power in the output of an FFT filter times the input data sampling rate. Every signal contains some noise. That noise is generally spread over the frequency spectrum of interest, and each narrowband filter passes a certain amount of that noise through its main lobe and sidelobes. White noise is used as the input signal and the noise power out of each filter is compared to the noise power into the filter to determine the equivalent noise bandwidth of each passband filter. In other words, equivalent noise bandwidth represents how much noise would come through the filter if it had an absolutely flat passband gain and no sidelobes. It should be noted that leakage to the side lobes not only decreases the magnitude of the main lobe but also increases the frequency resolution (spreading) of the main lobe. The width of the spread is compared to the rectangular window, which has a main lobe width of 1. Table 16.1 shows the equivalent bandwidth spread to be greater than 1 for all window functions.

16.2.6 Three-Decibel Main-Lobe Bandwidth

The standard definition of a filter's bandwidth is the frequency range over which sine waves can pass through the filter without being attenuated more than a factor of 2 (3 dB) relative to the gain of the filter at its center frequency. The narrower the main lobe, the smaller the range of frequencies that can contribute to the output of any FFT filter. This means that the accuracy of the FFT filter, in defining the frequencies in a waveform, is improved by having a narrower main lobe.

16.3   WINDOW FUNCTIONS AND WEIGHTING EQUATIONS

16.3.1 The Rectangle Window

In this section, some of the most used window models to reduce spectral leakage are presented. Some terms will be used to relate the effects of the particular window function to the rectangular window. Recall that the reduction of the side-lobe leakage introduces leakage from the expansion of the main lobe. Also some gain is lost because of the main

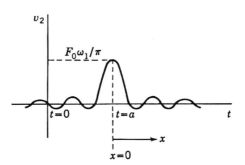

FIGURE 16.5: The Sinc function. Note the ripples

lobe spreading. More noise is introduced in the spectrum, which is a characteristic that is most often forgotten. So the question an engineer must consider is "What is the best solution or what is the best window?" According to the characteristics of the data, the analyst must select a function that produces the least error and complications.

Let us begin with the most common window, which is the "Rectangular window." The simplest way to model a finite record length is through the usage of the rectangular function (often called boxcar). The function equals one in the sampling interval and it is zero everywhere else. The Window Weighting function is given by (16.1).

$$w(n) = 1; \quad \text{For} \quad n = 0 \ to \ N - 1 \tag{16.1}$$

Consider a signal, $x(t)$, with a Fourier Transform, $X(f)$, and that the signal is defined in some finite period of time zero (0) to $+T$. The observed signal is multiplied by the window Weighting Function $w(t)$ in the time domain. The resulting function is the product of $w(t)$ and $x(t)$. The major limitation of the new function is that it is zero outside the time interval (observation window).

The Fourier transform of the window function $w(t)$ is the well-known *SINC* function. The *SINC(x)* is defined as $\sin(x)/x$, which is given in Fig. 16.5.

16.3.2 The Rectangular Window and the Gibbs Phenomenon

The rectangular function truncates a signal in a discontinuous manner at its edges. The Fourier transform approximates the function with a linear combination of sines and cosines. The Gibbs phenomenon expresses the fact that the *SINC* function oscillation is always present when discontinuities exist. If the discontinuity is smoothed with a window function, the fit will suppress the leakage, which will result in a smaller error. As the number of linear combinations in the transformation approaches infinity, the MSE in the region of the discontinuity approaches a steady error value of 0.09 (9%).

When the Fourier coefficients are squared, the negative parts of the waveform become positive. In the spectrum, the squared ripples of the Sinc function are the sidelobes. One should note that in Fig. 16.5 the following are true:

1. The SINC function is centered at the main lobe frequency "*MAXIMA*," which is an indication that the majority of the information remains around the fundamental frequency (f_0 Hz) or its multiples (harmonics).

2. The transformed function $X(f)$ is zero at the crossover points, which introduces "SMEARING." Instead of having information centered on 0, the information is centered in a RANGE around f_0.

The spectrum of the Rectangular Window function is shown in Fig. 16.6.

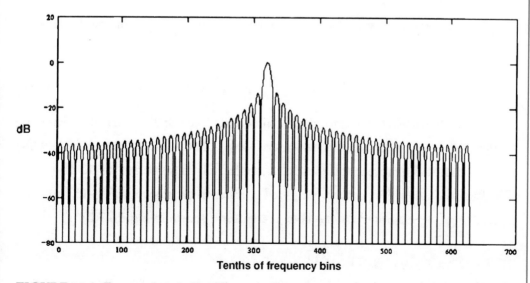

FIGURE 16.6: Rectangular window. The trace shows the main frequency lobe and the resulting side lobes. The first side lobe is -13 dB from the peak of the frequency main lobe

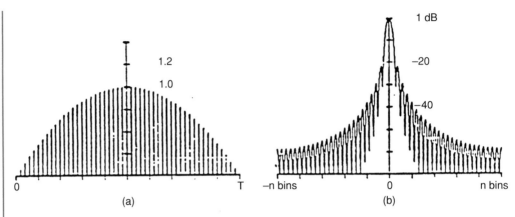

FIGURE 16.7: Parzen/Rietz window. Trace (a) is the time domain weighted coefficients. Trace (b) shows the main frequency lobe and the resulting side lobes. The first sidelobe is −21 dBs from the peak of the frequency main lobe

Characteristics of the rectangular window performance are −13 dB of the first side lobe from the height of the main lobe with a −6 dB/octave fall-off rate, 1.00 coherence gain, and 3.92-dB scallop loss (3.92-dB worst case process loss).

16.3.3 Parzen/Rietz Window

The Parzen/Rietz window is the simplest continuous polynomial window. The mathematical model is shown in (16.2).

$$x(i) = 1 - \left| \frac{(i - 0.5(N-1))}{0.5(N-1)} \right| \tag{16.2}$$

The Parzen/Rietz window gives a smoother representation than the Hamming window; however, the first side lobe decrease is only −21 dB. The window exhibits a discontinuous first derivative at the boundaries, so the roll-off per octave is −12 dB/octave. The Parzen window time domain weighted coefficients are shown in Fig. 16.7(a), with its resulting Log-magnitude of the Fourier transform shown in Fig. 16.7(b).

16.3.4 The Tukey Family of Windows

The Tukey window can be described as a cosine lobe of width, (alpha/2)(N), convolved with a rectangle window of width, (1.0 to alpha/2)/N. The term *alpha* represents a parameter, and according to this value we have a family of windows. The usual value of alpha is 0.10 or "10% Tukey Window" (often called a ten percent window) [1]. Another name for the family of these windows is cosine-tapered windows. The main idea behind any window is to smoothly set the data to zero at the boundaries, and yet minimize the reduction of the process gain of the windowed transform. There are several Tukey windows for different values of alpha ranging from 0.10 to 0.75.

The mathematical model for the ten percent Tukey window is given by equations in (16.3). Since the formula is taken from our signal processing program, the variable "alpha" has a value of 0.10.

$$w(i) = 0.5 \left(1 - \cos \left(\frac{10\pi i}{10} \right) \right) \quad \text{for } i = 00 \text{ to } 0.1N$$

$$w(i) = 1 \quad \text{for } i \text{ between } i = 0.1N \text{ to } 0.9N \qquad (16.3)$$

$$w(i) = 0.5 \left(1 - \cos \left(\frac{10\pi i}{10} \right) \right) \quad \text{for } i = 0.9N \text{ to } N$$

Notice that for the last and first ten percent of data the window function is the same. Higher alpha values tend to increase the main lobe spread with less leakage results. For an alpha of 0.10, the Tukey window has the first side-lobe level at -14 dB with a roll-off rate of -18 dB. It is interesting that the first side-lobe level and the roll-off rate characteristics remain the same for alpha values up to 0.25. The Tukey window time domain weighted coefficients for alpha values of 0.25, 0.5, and 0.75 are shown in Figs. 16.8(a), 16.9(a), and 16.10(a), respectively. The resulting Log-magnitude of the Fourier transform for alpha values of 0.25, 0.5, and 0.75 are shown in Figs. 16.8(b), 16.9(b), and 16.10(b), respectively [1] and [2].

16.3.5 Hanning Windows

The Hanning family of windows is also known as the "Cosine Windows," since the functions used are true cosines. For the readers, historical information, the name "Hanning" does not exist, since the founder of these functions was an Austrian meteorologist named "Hann." The *ing* ending was added because of the popularity of the cosine window models. The mathematical model for the Hanning window is given by equations in (16.4).

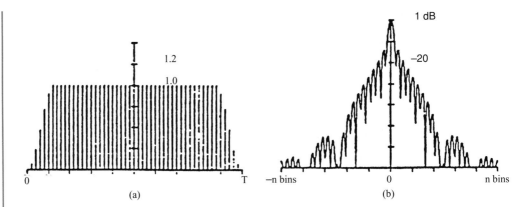

(a)   (b)

FIGURE 16.8: Twenty-five percent (25%) Tukey window. For alpha of 0.25, the trace (a) is the time domain weighted coefficients, whereas the trace (b) shows the main frequency lobe and the resulting side lobes. The first side lobe is −14 dBs from the peak of the frequency mainlobe with a −18-dB per octave side lobe fall-off rate

The variable "alpha" is an exponent of the cosine with a range from 1 to 4.

$$w(i) = 1 - \cos^{alpha}(x) \qquad (16.4)$$

As alpha takes larger values, the windows are becoming smoother and smoother, and the first side lobe is smaller. The family of Hanning window time domain weighted

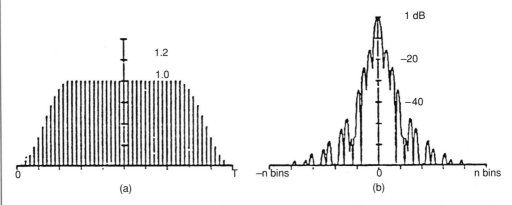

(a)   (b)

FIGURE 16.9: Fifty percent (50%) Tukey window. For alpha of 0.50, the trace (a) is the time domain weighted coefficients, whereas the trace (b) shows the main frequency lobe and the resulting side lobes. The first side lobe is −15 dBs from the peak of the frequency main lobe with a −18 dB per octave sidelobe fall-off rate

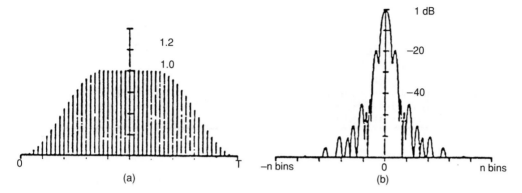

FIGURE 16.10: Seventy-five percent Tukey window. For alpha of 0.75, the trace (a) is the time domain weighted coefficients, whereas the trace (b) shows the main frequency lobe with the first side lobe at −19 dB and a −18 dB per octave side lobe fall-off rate

coefficients for alpha values of 1, 2, 3, and 4 are shown in Figs. 16.11(a), 16.12(a), 16.13(a), and 16.14(a), respectively. The resulting Log-magnitude of the Fourier transform for alpha values of 1, 2, 3, and 4 are shown in Figs. 16.11(b), 16.12(b), 16.13(b), and 16.14(b), respectively. Notice in the figures what effect increasing the value of alpha from 1 to 4 has on the first side-lobe value and the side-lobe roll-off rate. Note that for each increase of alpha by 1 the absolute magnitude of the side-lobe roll-off rate increases by 6 dB.

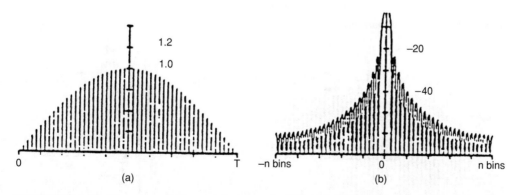

FIGURE 16.11: Hanning window with alpha of 1. The trace (a) is the time domain weighted coefficients, whereas the trace (b) shows the main frequency lobe with the first side-lobe at −23 dB and a −12 dB per octave side-lobe fall-off rate

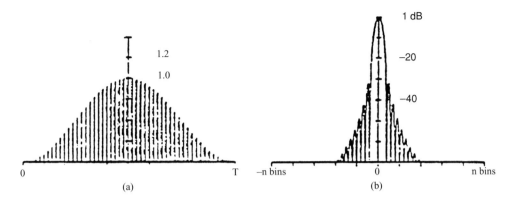

(a)   (b)

FIGURE 16.12: Hanning window with Alpha of 2. The trace (a) is the time domain weighted coefficients, whereas the trace (b) shows the main frequency lobe with the first side lobe at -32 dB and a -18 dB per octave side-lobe fall-off rate

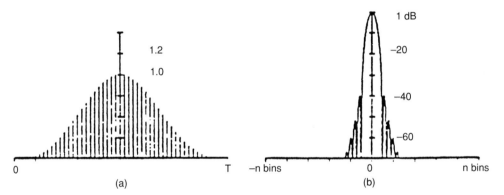

(a)   (b)

FIGURE 16.13: Hanning window with alpha of 3. The trace (a) is the time domain weighted coefficients, whereas the trace (b) shows the main frequency lobe with the first side lobe at -39 dB and a -24 dB per octave side-lobe fall-off rate

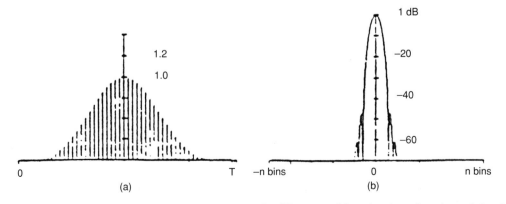

(a)   (b)

FIGURE 16.14: Hanning window with alpha of 4. The trace (a) is the time domain weighted coefficients, whereas the trace (b) shows the main frequency lobe with the first side lobe at -47 dB and a -30 dB per octave side-lobe fall-off rate

From trigonometric identities, an alternate method for expressing the cosine window is to write the expression in terms of the "Sine" function, when the data are not symmetrical about the origin. For example if $\alpha = 2$ the cosine model becomes as shown in by (16.5).

$$\sin^2(x) = 1 - \cos^2(x) \tag{16.5}$$

Hence, the basic cosine model could be expressed as (16.6).

$$w(n) = \sin^{\text{alpha}}\left(\frac{n\pi}{N}\right) \text{ for alpha values of 2 to 4} \tag{16.6}$$

16.3.6 Welch Window

This window is a modification of the Parzen window with some better characteristics especially in the coherent gain; however, the Welch window produces greater main lobe leakage due to a wider spreading. The mathematical model for the Welch window is given by (16.7). Notice the polynomial similarity between Parzen and Welch expressions.

$$w(i) = 1 - \left(\frac{i - 0.5(N-1)}{0.5(N-1)}\right)^2 \tag{16.7}$$

This type of window is not good for signals with closely related spectral components. The main side-lobe effect can be worse than the leakage from the rectangular window.

16.3.7 Blackman Window

The name "Blackman Window" was given to honor Dr. Blackman, a pioneer in the harmonic signal detection. There are several windows under his name, differing in the amount of variables used. For simplicity only the "Exact Blackman Window" will be described, since this window has superb tone discrimination characteristics (used in acoustics).

The expression for the Exact Blackman Window is given by (16.8).

$$w(n) = 0.42 + 0.50 \cos\left[\frac{2\pi}{N}n\right] + 0.08\left[\frac{2\pi}{N}2n\right] \tag{16.8}$$

where $n = -N/2, \ldots, -1, 0, +1, \ldots, +N/2$

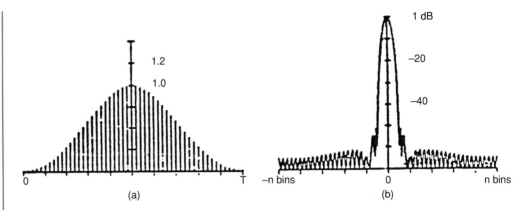

FIGURE 16.15: The Exact Blackman Window. Trace (a) is the time domain weighted coeffi-cients. Trace (b) shows the main frequency lobe and the resulting side lobes. The first sidelobe is -51 dB from the peak of the frequency main lobe

Notice that the third term deals with the second harmonic. If we add the terms at $n = 0$ then $w(n) = 0$, something that we want to see. The side-lobe level of the Exact Blackman Window is -51 dB almost 3.5 times lower than the normal rectan-gular window. The roll-off for the Exact Blackman Window is -18 dB/octave. These characteristics are impressive, since the second side-lobe is at a level of -69 dB. What these numbers indicate is that the second harmonic leakage is about 1200 times less than the maximum value at the center frequency f_0. For a normal power spectrum, most analysts are not interested below -60 dB; hence, a value of -69 dB is exceptionally good (the decibel values are referred to normalized spectra). The disadvantage of the Exact Blackman Window is the coherent gain, which unfortunately is less than 0.5, closer to 0.47. However, this error can be compensated in the frequency domain by multiplying with the inverse of the coherent gain.

The Exact Blackman Window time domain weighted coefficients are shown in Fig. 16.15(a) with its resulting Log-magnitude of the Fourier transform shown in Fig. 16.15(b).

16.4   SUMMARY

Leakage is the result of making an infinitely long function finite in length. A smearing effect is the result of convolving the infinite period of the signal, with the SINC function.

The effects of leakage can be suppressed or reduced with the application of special functions called "Windows." As many as 58 window functions have been developed with various optimizations in mind. In practice, most engineers will apply a Hanning or Hamming window; however, it is not recommended that only one window be used for all applications, but rather that several windows should be tried for the best results in the particular situation.

16.5 REFERENCES

1. Bendat, J., and Piersol, A. *Randon Data Analysis and Measurement Procedures*, 2nd Edition, Wiley, pp. 393–400.

2. Gabel, R., and Roberts, R. *Signals and Linear Systems*, pp. 253–370.

3. Flannery, B., Tenkolsky, S., and Veterling, W. *Numerical Recipes: The Art of Scientific Computing*. Wesley Press, pp. 381–429.

4. Van Valkenburg, M. E. *Network Analysis*, pp. 452–495.

16.6 SUGGESTED READING

1. Bergland, G. D. "A Guided Tour of the Fast Fourier Transform," IEEE Spectrum, pp. 41–52, July 1969.

2. Harris, F. "The use of Windows for Harmonic Analysis with DFT," Proceedings of IEEE, Vol. 6, No. 1, January 1978.

3. Geckinki, N., and Yanuz, D. *IEEE Transactions on Acoustics Speech and Signal Processing*. Vol. 26, No. 6, December 1978.

4. Nudall, A. "Some Windows With Very Good Sidelobe Behavior," IEEE Transactions on Acoustics Speech and Signal Processing, Vol. 29, No. 1, February 1981.

CHAPTER 17

Transfer Function Via Spectral Analysis

17.1 INTRODUCTION

This chapter describes the method for estimation of the frequency response of a linear system, using spectral density functions. The autocorrelation and cross-correlation functions are discussed to provide background for the development of their frequency domain counterparts, the autospectral and cross-spectral density functions. The coherence function is also discussed as a means of calculating the error in the transfer function estimation. Applications will be restricted to single input/single output models, where it is assumed that systems are constant-parameter, linear types.

To meet control system specifications, a designer needs to know the time response of the controlled variable in the system. Deriving and solving the differential equations of the system can obtain an accurate description of the system's response characteristics; however, this method is not particularly practical for anything but very simple systems. With this solution, it is very difficult to go back and decide what parameters need to be adjusted if the performance specifications are not met. For this reason, the engineer would like to be able to predict the performance of the system without solving its associated differential equations and he/she would like for his/her analysis to show which parameters need to be adjusted or altered in order to meet the desired performance characteristics.

The transfer function defines the operation of a linear system and can be expressed as the ratio of the output variable to the input variable as shown in Fig. 17.1.

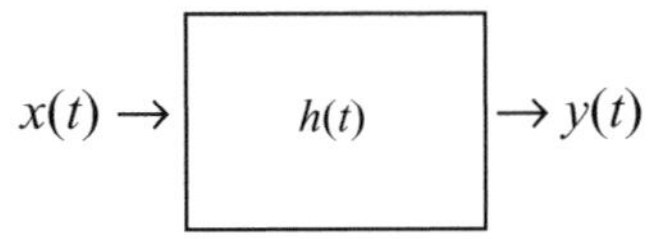

FIGURE 17.1: Transfer function from input–output

The common approach to system analysis is to determine the system's transfer function through observation of its response to a predetermined excitation signal. Evaluation of the system characteristics can be performed in the time or frequency domains.

17.2   METHODS

The engineer has several methods available to him/her for evaluating the system's response. The Root Locus Method is a graphical approach in the time domain, which gives an indication of the effect of adjustment based on the relation between the poles of the transient response function, and the poles and zeros of the open loop transfer function. The roots of the characteristic equation are obtained directly to give a complete solution for the time response of the system.

In many cases it is advantageous to obtain system performance characteristics in terms of response at particular frequencies. Two graphical representations for transfer function derivations are the Bode and Nyquist Methods. The Bode Plot Method yields a plot of the magnitude of the output/input ratio versus frequency in rectangular or logarithmic coordinates. A plot of the corresponding phase angle versus frequency is also produced. The Nyquist Plot Method is very similar: it plots the output/input ratio against frequency in polar coordinates. The amplitude and phase data produced with these methods are very useful for obtaining an estimate of the system's transfer function. They can be determined experimentally for a steady-state sinusoidal input at various frequencies, from which the magnitude and phase angle diagrams are then derived directly, leading to the synthesis of the transfer function.

Analytical techniques that fit rational polynomial transfer functions using digital computers are very popular as they speed the evaluation process considerably. In the next section, two functions, which provide analytical methods for measuring the time-domain properties of signal waveforms, will be developed.

17.3 AUTOCORRELATION

The autocorrelation function gives an average measure of the time-domain properties of a signal waveform. It is defined as (17.1):

$$R_{xx}(\tau) = \lim_{T_0 \to \infty} \frac{1}{T_0} \int_{-T_0/2}^{T_0/2} f(t) f(t + \tau)\, dt \tag{17.1}$$

This is the average product of the signal, $f(t)$, and a time-shifted version of itself, $f(t + \tau)$. The expression above applies to the case of a continuous signal of infinite duration. In practice, the intervals must be finite and it is necessary to use a modified version as given by (17.2).

$$R_{xx}(\tau) = \int_{-\infty}^{\infty} f_1(t)\, f_1(t + \tau)\, dt \tag{17.2}$$

The autocorrelation function may be applied to deterministic as well as random signals. Each of the frequency components in the signal $f(t)$ produces a corresponding term in the autocorrelation function having the same period in the time-shifted variable, τ, as the original component has in the time variable, t. The amplitude is equal to half of the squared value of the original. The phase shifts of each signal produce a single cosine term in the autocorrelation as shown in Fig. 17.2.

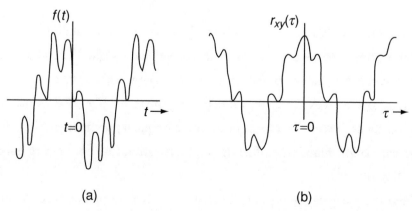

(a)   (b)

FIGURE 17.2: Autocorrelation. Trace (b) is the autocorrelation of the signal shown in (a)

The fact that the autocorrelation is composed of cosines implies that it must be an even function of τ. When $\tau = 0$, all of the cosine functions reinforce each other to give a peak positive value for the autocorrelation. Whether this peak value is ever attained for other shifts of τ depends on whether the components of the signal are harmonically related or not. The peak value of the autocorrelation function at $\tau = 0$ is simply the mean square value, or average power of the signal, and is given by (17.3) [1].

$$R_{xx}(\tau)\Big|_{\tau = 0} = R_{xx}(0) = \lim_{T_0 \to \infty} \int_{-T_0/2}^{T_0/2} [f(t)]^2 \, dt \tag{17.3}$$

17.4   THE CROSS-CORRELATION FUNCTION

The cross-correlation function is essentially a time-averaged measure of shared signal properties and is defined as (17.4).

$$R_{xy}(\tau) = \lim_{T_0 \to \infty} \frac{1}{T_0} \int_{-T_0/2}^{T_0/2} f_1(t) f_2(t + \tau) \, dt \tag{17.4}$$

where τ is a time shift on one of the signals. Since signals to be compared must be of finite duration, the function is modified as shown in (17.5).

$$R_{xy}(\tau) = \int_{-\infty}^{\infty} f_1(t) \, f_2(t + \tau) \, dt \tag{17.5}$$

The cross-correlation function reflects the product of the amplitudes of $f_1(t)$ and $f_2(t)$, their common frequency, ω, and their relative phase angle. When the two signals being cross-correlated share a number of common frequencies, each gives a corresponding contribution to the cross-correlation function as shown in Fig. 17.3. Since it retains information about the relative phases of common frequency components in two signals, the cross-correlation function, unlike the autocorrelation function, is not normally an even function of τ (4).

From the previous discussions, it can be seen that the autocorrelation and cross-correlation functions provide all analytical means, which may evaluate the magnitude

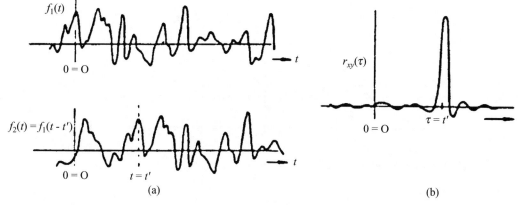

FIGURE 17.3: Cross-correlation function. Trace (b) is the cross-spectral estimate of the two traces in (a). The bottom trace in (a) is the time shifted upper trace

and phase of a system's time response. The next step is to look at the equivalents of these functions in the frequency domain where computation will be simplified.

17.5   SPECTRAL DENSITY FUNCTIONS

Spectral density functions can be derived in several ways. One method takes the direct Fourier transform of previously calculated autocorrelation and cross-correlation functions to yield the two-sided spectral density functions given in (17.6).

$$S_{xx}(f) = \int_{-\infty}^{\infty} R_{xx}(\tau)\, e^{-j2\pi f\tau}\, d\tau$$

$$S_{yy}(f) = \int_{-\infty}^{\infty} R_{yy}(\tau)\, e^{-j2\pi f\tau}\, d\tau$$

$$S_{xy}(f) = \int_{-\infty}^{\infty} R_{xy}(\tau)\, e^{-j2\pi f\tau}\, d\tau \qquad (17.6)$$

These integrals always exist for finite intervals. The quantities $S_{xx}(f)$ and $S_{yy}(f)$ are the autospectral density functions of signals $x(t)$ and $y(t)$ respectively, and $S_{xy}(f)$ is the cross-spectral density function between $x(y)$ and $y(t)$. These quantities are related

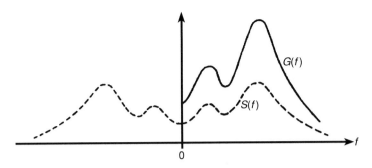

FIGURE 17.4: One- and two-sided autospectrum

as in (17.7).

$$S_{yy}(f) = \left| H(f) \right|^2 S_{xx}(f)$$
$$S_{xy}(f) = H(f)\, S_{xx}(f) \tag{17.7}$$

In terms of physically measurable one-sided spectral density functions where f varies over $(0, \infty)$, the results are given in (17.8):

$$G_{xx}(f) = 2 S_{xx}(f)$$
$$G_{yy}(f) = 2 S_{yy}(f)$$
$$G_{xy}(f) = 2 S_{xy}(f) \tag{17.8}$$

Figure 17.4 shows the relation between the two-sided autospectral density function and the one-sided representation.

The autospectral and cross-spectral density functions are also shown in (17.9).

$$G_{yy}(f) = \left| H(f) \right|^2 G_{xx}(f)$$
$$G_{xy}(f) = H(f)\, G_{xx}(f) \tag{17.9}$$

Note that the top equation of (17.9) is a real-valued relation which contains only the transfer function gain factor of the system, $|H(f)|$. The lower equation is a complex-valued relation which can be broken down into a pair of equations to give both the gain factor, $|H(f)|$, and the transfer function phase factor, $\square\,(f)$ of the system. Recall that the FFT of the cross-correlation yields the complex cross-spectra as given by the following equation. Note that the cross-spectrum is a complex function with real and imaginary functions, where $C_{xy}(f)$ is the coincident spectral density function (cospectrum) and $Q_{xy}(f)$ is the quadrature spectral density function (quad-spectrum) as shown in (17.10).

$$G_{xy}(f) = 2 \int R_{xy}(\tau)\, e^{-j2\pi f\tau}\, d\tau = C_{xy}(f) - jQ_{xy}(f) \tag{17.10}$$

In complex polar notation, the cross-spectral density becomes 17.11.

$$G_{xy}(f) = |G_{xy}(f)|\, e^{-j\theta_{xy}(f)}$$
$$H(f) = |H(f)|\, e^{-j\phi(f)} \tag{17.11}$$

where

$$|G_{xy}(f)| = (C_{xy}^2(f) + Q_{xy}^2(f))^{1/2}$$
$$\theta_{xy}(f) = \tan^{-1}\left(\frac{Q_{xy}(f)}{C_{xy}(f)}\right)$$

Thus, by knowing the autospectra of the input $G_{xx}(f)$ and the cross-spectra $G_{xy}(f)$ one can determine the transfer function, $|H(f)|$. However, the magnitude of the transfer function does not provide any phase information. Nevertheless, the complete frequency response function with gain and phase can be obtained when both $G_{xx}(f)$ and $G_{xy}(f)$ are known [2]. Figure 17.5 illustrates these relationships.

An alternative direct transform method is widely used and does not require computation of the autocorrelation and cross-correlation functions beforehand. It is based on finite Fourier transforms of the original data records. For any long, but finite records of length, T, the frequency domain equation is written as (17.12).

$$Y(f) = H(f)X(f) \tag{17.12}$$

where $X(f)$ and $Y(f)$ are finite Fourier transforms of $x(t)$ and $y(t)$ respectively. It follows that (17.12) can be rewritten as (17.13).

$$Y^*(f) = H^*(f)X^*(f)$$
$$|Y(f)|^2 = |H(f)|^2\,|X(f)|^2$$
$$X^*(f)Y(f) = H(f)|X(f)|^2 \tag{17.13}$$

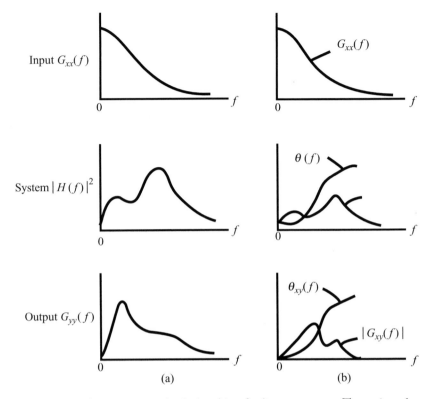

FIGURE 17.5: Input/output spectral relationships for linear systems. Traces in column (a) are autospectra while traces in column (b) are cross-spectra

With some algebraic manipulation, (17.14) can be obtained.

$$G_{yy}(f) = |H(f)|^2 G_{xx}(f)$$
$$G_{xy}(f) = H(f) G_{xx}(f) \tag{17.14}$$

The complex conjugation of the lower equation yields (17.15).

$$G_{xy}^*(f) = G_{yx}(f) = H^*(f) G_{xx}(f) \tag{17.15}$$

where

$$G_{xy}(f) = |G_{xy}(f)| e^{j\theta_{xy}(f)}$$
$$H^*(f) = |H(f)| e^{j\phi(f)}$$

Thus, the phase factor can be determined from (17.16).

$$\frac{G_{xy}(f)}{G_{yx}(f)} = \frac{H(f)}{H^*(f)} = e^{-j2\phi(f)} \tag{17.16}$$

and the complete response function becomes (17.17).

$$G_{yy}(f) = H(f)[H^*(f)G_{xx}(f)] = H(f)G_{yx}(f) \tag{17.17}$$

For the ideal single input/single output model, the frequency response is given by (17.18).

$$H(f) = \frac{G_{xy}(f)}{G_{xx}(f)} \tag{17.18}$$

The magnitude of the transfer function is commonly expressed in logarithmic form, as decibels. The logarithmic of the magnitude of the transfer function $G(f)$ expressed in decibels as (17.19).

$$20\log|H(f)| \quad \text{in} \quad \text{dB} \tag{17.19}$$

Since the transfer function is a function of frequency, the log magnitude is also a function of frequency [1 and 3].

17.6   THE COHERENCE FUNCTION

System error is defined as the ideal or desired system output value minus the actual system output. The ideal value establishes the desired performance of the system. Coherence provides a measure of the linear correlation between two components of a process at a particular frequency. The coherence function between the input $x(t)$ and the output $y(t)$ is a real-valued quantity defined by (17.20).

$$\gamma_{xy}^2(f) = \frac{|G_{xy}(f)|^2}{G_{xx}(f)} G_{yy}(f) \tag{17.20}$$

where

$$0 \le \gamma_{xy}^2 \le 1 \quad \text{for all} \quad f$$

For a constant parameter linear system with a single, clearly defined input and output, the coherence function will be unity. If $x(t)$ and $y(t)$ are completely unrelated, the coherence function will be zero. Coherence function values greater than 0 but less than 1 may be due to the presence of noise in the measurements, or there may be nonlinearities in the system relating $x(t)$ and $y(t)$, or it may be that the output $y(t)$ may be a function of or related to other endogenous inputs in addition to the input signal $x(t)$. For linear systems, the coherence function can be interpreted as the fractional portion of the mean square value at the output $y(t)$ that is contributed by $x(t)$ at frequency f. Conversely, the quantity $[1 - \gamma_{xy}^2(f)]$ is a measure of the mean square value of $y(t)$ not accounted for by $x(t)$ at frequency f.

17.7   SUMMARY

The ability to derive the transfer function for a system's response is an integral part of the design process. Of the graphical and analytical techniques available to the engineer, the spectral density functions prove to be powerful computational tools for this purpose. In addition, through the coherence function, the spectral density functions provide an effective means for evaluating error in the transfer function model.

17.8   REFERENCES

1. Bendat, J. S., and Piersol, A. G., *Random Data: Analysis and Measurement Procedures*, 2nd Edition, John Wiley and Sones, 1986.

2. D'Azzo, J. J., and Houpis, C., *Linear Control System Analysis and Design: Conventional and Modern*. McGraw-Hill, 1981.

3. Lessard, C. S. "Analysis of Nystagmus Response to Pseudorandom Velocity $\tau = 0$ Input," In *Computer Methods and Programs in Biomedicine*, Vol. 23, 1986, pp. 11–18.doi:10.1016/0169-2607(86)90075-1

4. Lynn, P. A. *An Introduction to the Analysis and Processing of Signals*. Halsted Press, 1973.

CHAPTER 18

Coherence Function from Spectral Analysis

18.1 INTRODUCTION

The coherence function is a mathematical tool used to determine the cause-and-effect relationship between the input and the output of a system. The coherence function is used in modeling and transfer function estimation to determine the "goodness fit" of the model or transfer function to the data. In addition, the coherence function is used to determine the "random error" of the estimates of functions which use spectral analysis. Several examples given and a description of the coherence function in the context of a correlation coefficient is included for clarity. Like all scientific tools, the coherence function has its limitations and biases. These are explained in general, and specific references are cited. Two experimental examples are given showing that the coherence function can be used in primary analysis of an input–output system or at the end of a series of steps used in analyzing a more sophisticated type input–output system.

Signal-processing problems involve determining how various signals are modified in amplitude, phase, or frequency as the signals pass through some type of system. The system may be an airplane, various electromechanical devices, machine, or the human body. The block diagram model in Fig. 18.1 shows how a signal may be altered by a system.

In Fig. 18.1, the system may modify the input signal so that the output signal $Y(f)$ has different characteristics than the input signal $X(f)$. Subsystems may be connected

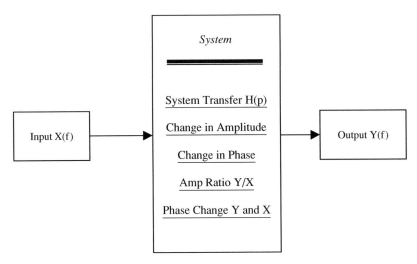

FIGURE 18.1: System modification of input signal

to other subsystems in such a way that each subsystem will modify its respective input such that the total output is changed. In addition, the system itself may generate some type of noise or interference that is only an artifact, which has nothing to do with the actual function of the system. Noise may cause alteration in the output which may make the output appear as having been modified by the system. Another type of system may have multiple inputs as illustrated in Fig. 18.2.

In systems with multiple inputs, one would be evaluating causability. The signal-processing objective is to determine how much each source of the input signal though the system transfer, $H(f)$, is responsible for the output. Other types of systems involve

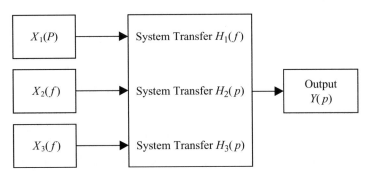

FIGURE 18.2: Multiple input, single output system

multiple inputs with multiple output systems or single input with multiple outputs systems [1]. The relationships between the input transfer function of the system and the output are analyzed by several techniques [2].

This chapter will deal primarily with coherence function of single input, single output systems only. The coherence function is a function of frequency. Its maximum values occur at the frequencies where the greatest transfer of energy within a system takes place.

18.2 DESCRIPTION OF THE COHERENCE FUNCTION

The simplest explanation of the coherence function would be that it is similar to the correlation coefficient, r[3]. The correlation coefficient, r, is a measure of the linear relationship between two variables, with one being the independent variable and the other being the dependent variable. Mathematically, the correlation coefficient, r, is shown in terms of variance as given by (18.1) through (18.5), where (18.1) through (18.3) are the mathematical definitions of terms.

$$\sigma_X^2 = \frac{1}{n} \sum_{i=1}^{n} Xi\, Xi = \text{Variance of } X \tag{18.1}$$

$$\sigma_y^2 = \frac{1}{n} \sum Yi\, Yi = \text{Variance of } Y \tag{18.2}$$

$$yx = \frac{1}{n} \sum Yi\, Xi = \text{Covariance of } X \text{ and } Y \tag{18.3}$$

$$r_{xy} = \hat{\rho}_{xy} = \frac{S_{xy}}{S_x S_y} = \frac{\sum_{t-1}^{N} (x_t - \bar{x})(y_t - \bar{y})}{\left[\sum_{t-1}^{N} (x_t - \bar{x})^2 \sum_{t-1}^{N} (y_t - \bar{y})^2\right]^{1/2}} \tag{18.4}$$

or

$$= \frac{\sum_{t-1}^{N} x_t y_t - N\bar{x}\,\bar{y}}{\left[\left(\sum_{t-1}^{N} x_t^2 - N\bar{x}^2\right)\left(\sum_{t-1}^{N} y_t^2 - N\bar{y}^2\right)\right]^{1/2}} \tag{18.5}$$

The correlation coefficient, r, value has a range from -1 to $+1$. The correlation coefficient squared term r^2 is the coefficient of determination, which is often referred to as the "Goodness Fit." The correlation coefficient squared shows the percentage of variation in the output or dependent variable that is accounted for from the input or independent variable. One word of caution when using this comparison is the correlation coefficient squared r^2 between one dependent and multiple independent variables is that the r^2 does not represent the linear relationship. There may be linear, quadratic, or cubic relationship contained within the correlation coefficient squared r^2 term.

The coherence function uses almost identical calculations when comparing two spectra, where

G_{xx} (of autospectrum 1) is analogous to (18.1),

G_{yy} (of autospectrum 2) is analogous to (18.2), and

G_{xy} (of both cross-spectrums) is analgous to (18.3).

The coherence function , γ^2, is equivalent to the correlation coefficient squared, r^2. It should be noted that the correlation coefficient, r^2 is a single value measure of "Goodness," the coherence is a function with values related to frequency analysis of continuous and/or sampled random data. Therefore, the coherence function is evaluated at specific analysis frequencies. From (18.6), it can be noted that it is necessary to calculate the autospectra of both the input and output signals, as well as, the cross-spectra of the input–output signals.

$$\gamma_{xy}^2(f) = \frac{|\hat{G}_{xy}(f)|^2}{\hat{G}_{xx}(f)\hat{G}_{yy}(f)} \tag{18.6}$$

In Fig. 18.1, a single input–output system is shown with transfer function $H(f)$. The coherence will show causality (unlike the correlation coefficient) between the input and the output. Another advantage over the correlation or cross correlation, which is used in signal processing, is that the coherence function is a function of frequency. The maximum value of one (1) for the coherence function occurs when all of the output (response) is directly attributed to the input (stimulus), whereas at lowest value, which is equal to 0, of the coherence function occurs at frequencies where little or no transfer of energy from input to output is taking place. If an output signal is interfered with by noise, the method of suppressing these depends on the frequencies at which they occur; therefore, the coherence function not only shows the causality between the input and the output but also may be of use in helping solve the interference or noise problems within a system.

In summary, the coherence is a frequency domain function with observed values ranging from 0 to 1. At each frequency where the coherence function is performed, it represents the fraction of the power output related to input. If the coherence function is less than 1, then there are three possible explanations:

1. there is noise in the system or

2. the system has some nonlinearity generating energy at other frequencies or

3. there are other inputs into the system that have not be accounted for.

The coherence function can be calculated using the Fast Fourier Transform [1]. If only one transform is performed on a power spectrum, the coherence function will always be unity. Therefore, it is necessary to average a number of transforms to get a good estimate of the function.

18.3   MISAPPLICATION OF THE COHERENCE FUNCTION

Let us look at an example of when a single coherence function by itself may not give desirable results to a noise problem. Assume that there are multiple inputs to a system as shown in Fig. 18.3.

The coherence function will be 1 if each suspect input is evaluated separately, while the other inputs are turned off. If two or more sources are related to or caused by the primary source, the coherence may not reveal the secondary cause, if both suspect inputs are wired to the same power source. This problem is similar to the "spurious correlation" and "multicolinearity" problems in statistics [4,7]. This problem may be solved by "multiple coherences" [6].

Bendat and Piersol [1] note there may be bias in the coherence function analysis of a frequency response, if there are other inputs correlated with the "input of interest," if

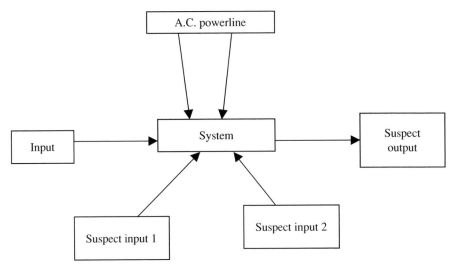

FIGURE 18.3: Spurious correlation in the analysis of a system

there is bias in the autospectral and cross-spectral density measurements, and if there are nonlinear and/or time-varying parameters within the system. Mathematical explanations of these biases are shown by (9.62) through (9.92) in Bendat and Piersol [1].

A confidence interval may be used with the coherence function. By definition, a confidence interval of 95%, for example, is "in repeated sampling 95% of all intervals will include the true fixed value of the parameter of interest" [4], then it is a true random interval.

18.4   EXAMPLES OF THE USE OF COHERENCE FUNCTIONS

The first example of how a coherence function might be used is demonstrated [5 and 6] to measure the vibration spectrum of a small printed circuit board shown in Fig. 18.4.

In this example, the ratio of acceleration/force was measured by connecting an accelerometer to the output and a pseudorandom signal to the input. In this case, the modal vibration resonances were measured and converted to the power spectrum. At 100 Hz, the coherence decreased from 1 to 0.83, indicating that the signal-to-noise ratio

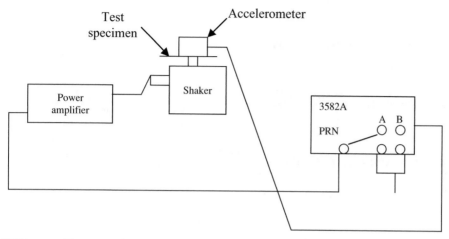

FIGURE 18.4: Test setup. The possible results in the evaluation would be (a) normal transfer function response or (b) transfer function is contaminated by vibration from loose hardware during testing of specimen

was $0.83/(1 - 0.83)$, so the coherence function is an alternative method of measuring the signal-to-noise ratio or an input–output system.

Lessard [3] used the coherence function to determine the cause-and-effect relationship between the output of the vestibular system and pseudorandom angular velocity input. The subject was subjected to pseudorandom angular velocity the saccade of the horizontal eye movement was converted to a signal. A full analysis of the output frequencies was performed. The coherence function analysis was performed as the last step in the analysis. This project was done with the objective of determining the nature and causes of space sickness.

18.5 PROBLEMS THAT LIMIT THE APPLICATION OF COHERENCE

There are numerous problems that limit the application. The major problems will be listed without discussion. Among these problems are

1. inaccurate source measurements,

2. source measurement interference,

3. correlation among sources,

4. input–output time-delay errors,

5. input measurement noise,

6. input–output nonlinearities,

7. input–output feedback,

8. reverberation effects, and

9. common source periodicities.

18.6 CONCLUSION

The coherence function, like many of the other mathematical operations used in signal processing, is susceptible to abuses by researchers who intentionally misapply its use or is not well informed with regard to its limitations. The coherence function should be used to establish a relationship between input and output of a system. None of the literature supports the idea of using the coherence function to compare two output signals such as the electromyograms (EMG) from two different muscles. Finally, as in all types of scientific tools, all assumptions must be met if the results are to be interpreted correctly.

18.7 REFERENCES

1. Bendat, J. S. and Piersol, A. G. *Random Data: Analysis and Measurement Procedures.* Wiley Interscience, 1971.

2. Bendat, J. S. "Statistical Errors in Measurement of the Coherence Function and Input–Output Quantities," *Journal of Sound and Vibration*, September 1978.

3. Lessard, C. S. "Analysis of Nystagmus Response to Pseudorandom Velocity Input," *Compute Methods and Programs in Biomedicine*, Vol. 23, 1986.

4. Neter, J., Wasserman, W., and Hunter M.H., *Applied Linear Regression Models.* Irwin, 1983.

5. Roth, P. R. "Measurements Using Digital Signal Analysis," *IEEE Spectrum*, April 1971.

6. Roth, P. R. "How to Use the Spectrum and Coherence Function," *Sound and Vibration*, January 1971.

7. Winer, B. J. *Statistical Principles in Experimental Design*. McGraw-Hill, 1971.

CHAPTER 19

Error in Random Data Estimate Analysis (Information Is Paraphrased from Bendat & Piersol)

This chapter presents the various statistical errors in random data analysis. Emphasis is on frequency domain properties of joint sample records from two different stationary (ergodic) random processes. The advanced parameter estimates discussed in this chapter include magnitude and phase estimates of autospectral density functions, cross-spectral density functions, transfer functions, and coherence functions. In particular, statistical error formulas are given without development or proof for the following functions:

1. frequency response function estimates (gain and phase),

2. coherence function estimates,

3. cross-spectral density function estimates, and

4. autospectral density function estimates.

19.1 CROSS-SPECTRAL DENSITY FUNCTION ESTIMATES

Recall that the cross-spectral density function between two stationary (ergodic) Gaussian random processes $\{x(t)\}$ and $\{y(t)\}$ is defined by (19.1). Specifically, given a pair of sample records $x(t)$ and $y(t)$ of unlimited length T, the one-sided cross-spectrum is given by (19.1).

$$G_{xy}(f) = \lim_{T \to \infty} \frac{2}{T} E\left[\bar{X}(f, T)Y(f, T)\right] \qquad (19.1)$$

where $X(f, T)$ is the conjugate of the input and $Y(f, T)$ are the finite Fourier transforms of $x(t)$ and $y(t)$, respectively.

It should be noted that the raw spectrum (no averages) of a finite length has the same equation as (19.1), without the limit as given by (19.2) and will have a resolution bandwidth given by (19.3).

$$\hat{G}_{xy}(f) = \frac{2}{T}[\bar{X}(f)Y(f)] \qquad (19.2)$$

$$B_e = \Delta f = \frac{1}{T} \qquad (19.3)$$

The ordinary coherence function is defined by (19.4).

$$\gamma_{xy}^2 = \frac{|G_{xy}|^2}{G_{xx}G_{yy}} \qquad (19.4)$$

The corresponding variance errors for "smooth" estimates of all of the "raw" estimates in (19.4) will be reduced by a factor of n_d when averages are taken over n_d statistically independent "raw" quantities. To be specific, the variance expressions are given by (19.5), (19.6), and (19.7).

$$\mathrm{Var}[\tilde{G}_{xx}] = \frac{G_{xx}^2}{n_d} \qquad (19.5)$$

$$\mathrm{Var}[\tilde{G}_{yy}] = \frac{G_{yy}^2}{n_d} \qquad (19.6)$$

$$\mathrm{Var}[|\tilde{G}_{xy}|] = \frac{|G_{xy}^2|}{\gamma_{xy}^2 n_d} \qquad (19.7)$$

Then, the normalized root-mean-square errors, which are the same as the normalized random errors, are given by (19.8), (19.9), and (19.10), respectively.

$$\varepsilon[\tilde{G}_{xx}] = \frac{1}{\sqrt{n_d}} \qquad (19.8)$$

$$\varepsilon[\tilde{G}_{yy}] = \frac{1}{\sqrt{n_d}} \qquad (19.9)$$

$$\varepsilon[|\tilde{G}_{xy}|] = \frac{1}{|\gamma_{xy}|\sqrt{n_d}} \qquad (19.10)$$

TABLE 19.1: Normalized Random Errors for Spectral Estimates

ESTIMATE	NORMALIZED RANDOM ERROR, ε
$\hat{G}_{xx}(f),\ \hat{G}_{yy}(f)$	$\dfrac{1}{\sqrt{n_d}}$
$\|\hat{G}_{xy}(f)\|$	$\dfrac{1}{\|\gamma_{xy}(f)\|\sqrt{n_d}}$
$\hat{C}_{xy}(f)$	$\dfrac{[G_{xx}(f)G_{yy}(f) + C_{xy}^2(f) - Q_{xy}^2(f)]^{1/2}}{C_{xy}(f)\sqrt{2n_d}}$
$\hat{Q}_{xy}(f)$	$\dfrac{[G_{xx}(f)G_{yy}(f) + Q_{xy}^2(f) - C_{xy}^2(f)]^{1/2}}{Q_{xy}(f)\sqrt{2n_d}}$

The quantity $|\gamma_{xy}|$ is the positive square root of γ_{xy}^2. Note that ε for the cross-spectrum magnitude estimate $|\hat{G}_{xy}|$ varies inversely with $|\gamma_{xy}|$ and approaches $\frac{1}{\sqrt{n_d}}$ as γ_{xy}^2 approaches 1.

A summary is given in Table 19.1 on the main normalized random error formulas for various spectral density estimates. The number of averages n_d represents n_d distinct (nonoverlapping) records, which are assumed to contain statistically different information from record to record. These records may occur by dividing a long stationary ergodic record into n_d parts, or they may occur by repeating an experiment n_d times under similar conditions. With the exception of autospectrum error estimation equation (19.8) and (19.9), all other error formulas are functions of frequency. Unknown true values of desired quantities are replaced by measured values when one applies these results to evaluate the random errors in actual measured data.

Let us work through an example to illustrate calculating the random errors in Frequency Response Transfer Function Estimate. Suppose the frequency response function between two random signals $x(t)$ and $y(t)$ is estimated using $n_d = 50$ averages. Assume that the coherence function at one frequency of interest is $\gamma_{xy}^2(f_1) = 0.10$ and at a second frequency of interest is $\gamma_{xy}^2(f_2) = 0.90$, the problem is to determine the random errors in the frequency response function gain and phase estimates at the two frequencies of interest.

TABLE 19.2: Multiple-Input/Output Model Estimates

FUNCTION BEING ESTIMATED	NORMALIZED RANDOM ERROR, ε
$\hat{\gamma}_{xy}^2(f)$	$\dfrac{\sqrt{2}[1 - \gamma_{xy}^2(f)]}{\lvert\gamma_{xy}(f)\rvert\sqrt{n_d}}$
$\hat{G}_{vv}(f) = \hat{\gamma}_{xy}^2(f)\hat{G}_{yy}(f)$	$\dfrac{[2 - \gamma_{xy}^2(f)]^{1/2}}{\lvert\gamma_{xy}(f)\rvert\sqrt{n_d}}$
$\lvert\hat{H}_{xy}(f)\rvert$	$\dfrac{[1 - \gamma_{xy}^2(f)]^{1/2}}{\lvert\gamma_{xy}(f)\rvert\sqrt{2n_d}}$
$\hat{\phi}_{xy}(f)$	$\text{S.D.}\,[\hat{\phi}_{xy}(f)] \approx \dfrac{[1 - \gamma_{xy}^2(f)]^{1/2}}{\lvert\gamma_{xy}(f)\rvert\sqrt{2n_d}}$

Equations (19.11) and (19.12) are used to calculate the variance and the normalized random error of the transfer function, respectively.

$$\text{Var}[\lvert\tilde{H}_{xy}\rvert] \approx \frac{(1 - \gamma_{xy}^2)\,\lvert H_{xy}\rvert^2}{2\gamma_{xy}^2 n_d} \qquad (19.11)$$

$$\varepsilon[\lvert\tilde{H}_{xy}\rvert] = \frac{\text{SD}\,[\lvert\tilde{H}_{xy}\rvert]}{\lvert H_{xy}\rvert} \approx \frac{(1 - \gamma_{xy}^2)^{1/2}}{\lvert\gamma_{xy}\rvert\,\sqrt{2n_d}} \qquad (19.12)$$

Then from (19.12), the normalized random error in the gain factor estimates at frequencies f_1 and f_2 will be given by (19.13) and (19.14).

$$\varepsilon[\lvert\tilde{H}_{xy}(f_1)\rvert] = \frac{(1 - 0.10)^{1/2}}{(0.32)(10)} = 0.30 \qquad (19.13)$$

$$\varepsilon[\lvert\tilde{H}_{xy}(f_2)\rvert] = \frac{(1 - 0.90)^{1/2}}{(0.95)(10)} = 0.033 \qquad (19.14)$$

The phase angle estimate, from the standard deviation (SD), is given by (19.15).

$$\text{SD}\,[\tilde{\phi}_{xy}] \approx \frac{(1 - \gamma_{xy}^2)^{1/2}}{\lvert\gamma_{xy}\rvert\,\sqrt{2n_d}} \approx \varepsilon\,\lvert\tilde{H}_{xy}\rvert \qquad (19.15)$$

Applying (19.15), the results also constitute approximations for the standard deviation (not normalized) in the phase factor estimate calculation as follows.

$$\text{SD}\,[\tilde{\phi}_{xy}(f_1)] = 0.30 \text{ radians} \qquad \text{SD}\,[\tilde{\phi}_{xy}(f_2)] = 0.033 \text{ radians}$$
$$= 17 \text{ degrees} \qquad\qquad = 1.9 \text{ degrees}$$

Hence, the estimates made with a coherence of $\gamma_{xy}^2(f_2) = 0.90$ are almost ten times as accurate as those made with a coherence of $\gamma_{xy}^2(f_1) = 0.10$.

19.2 SUMMARY

A summary is given in Table 19.2 on the main normalized random error formulas for single-input/output model estimates.

Biography

CHARLES S. LESSARD

Charles S. Lessard, Ph.D., Lt Colonel, United States Air Force (Retired), is an Associate Professor in the Department of Biomedical Engineering at Texas A&M University. His areas of specialization include *Physiological Signal Processing, Design of Virtual Medical Instrumentation, Noninvasive Physiological Measurements, Vital Signs, Nystagmus, Sleep & Performance Decrement, Spatial Disorientation, G-induced Loss of Consciousness (G-LOC),* and *Neural Network Analysis.* Dr. Lessard received a B.S. in Electrical Engineering from Texas A&M (1958) a M.S. from the U.S. Air Force Institute of Technology (1965), and his Ph.D. from Marquette University (1972).

As an officer in the U.S. Air Force, Lessard was a pilot of F86L Interceptors and B52G Strategic Bombers. He also served as Research Scientist and Chief of Biomedical Engineering Research for the Aerospace Medical Division of the School of Aerospace Medicine, at Brooks Air Force Base, Texas. In this capacity he planned and directed research efforts in biomedical projects associated with the Manned Orbiting Laboratory Program (MOL), developed medical instrumentation (EEG Analyzer), conducted research on computer on the analysis of sleep brainwaves and cardiac signals, and the effects of zero-gravity (0-G) on the cardiac response during valsalva maneuvers. At the U.S. Air Force Medical Research Laboratories, Wright-Patterson AFB, Lessard served with Biocybernetics Wing Engineering and worked on neural networks, self-organizing controls (SOC), and remotely piloted vehicles. He was the Field Office Director, Program Manager, with the Electronics Systems Division of the Air Force Systems Command during the installation and testing of Spain's Automated Air Defense System as a part of the Treaty of Friendship and Cooperation between the US and Spain. Dr. Lessard retired from the U.S. Air Force in 1981 after serving as the Deputy Director of the Bioengineering and Biodynamic Division at AMRL- Wright-Patterson Air Force Base. He began his academic career with Texas A&M University in 1981.

Charles Lessard was a Senior Scientist for Veridian Inc. at Brooks Air Force Base and lead scientist for KRUG Life Sciences, Inc. in the psychological and neuropsychological manifestations of spatial orientation, mechanisms of spatial orientation in flight, and countermeasures against spatial disorientation. Additionally, he was responsible for developing and conducting research in spatial disorientation and high acceleration (Gz-forces) induced loss of consciousness (G-LOC). He was a science and engineering expert for the USAF, Air Force Research Laboratories and Wyle Laboratories, Inc. on joint US military (Air Force and Navy) G-LOC studies performing analysis of physiological data, i.e., Auditory Evoked Responses (AER), electroencephalograms (EEG), electrocardiograms (ECG), electro-oculograms (EOG), Oxygen saturation (SaO_2), and Tracking Performance data.

Printed in the United States
by Baker & Taylor Publisher Services